Monographs on
Theoretical and Applied Genetics 1

Edited by
R. Frankel (Coordinating Editor), Bet-Dagan

M. Grossman, Urbana · H. F. Linskens, Nijmegen
D. de Zeeuw, Wageningen

J. Sybenga

Meiotic Configurations

A Source of Information for
Estimating Genetic Parameters

With 65 Figures

Springer-Verlag
Berlin Heidelberg New York 1975

Dr. J. SYBENGA
Department of Genetics of the
Agricultural University
Wageningen/Netherlands

ISBN 3-540-07347-7 Springer-Verlag Berlin Heidelberg New York
ISBN 0-387-07347-7 Springer-Verlag New York Heidelberg Berlin

Library of Congress Cataloging in Publication Data. Sybenga, Jacob, 1926-. Meiotic Configurations. (Monographs on theoretical and applied genetics; v. 1). Bibliography: p. Includes index. 1. Chromosomes. 2. Chromosome abnormalities. 3. Meiosis. I. Title. II. Series. QH600.S9. 574.8′732. 75-17562.

Typesetting and printing: Zechnersche Buchdruckerei, Speyer.
Bookbinding: Konrad Triltsch, Graphischer Betrieb, Würzburg.

Foreword

The aim of the monographs is to foster effective intra- and interdisciplinary communication between geneticists, and plant and animal breeders. This is to be achieved by publishing authoritative up-to-date texts; concise, but at the same time comprehensive, monographs, and multiauthor volumes on theoretical and applied genetics.

The following broad fields of genetics and breeding are within the scope of the series:

Evolutionary genetics	Developmental genetics
Population genetics	Biochemical genetics
Ecological genetics	Somatic cell genetics
Biometrical genetics	Agricultural genetics
Cytogenetics	Mutation breeding
Radiation genetics	Breeding methodology

Acceptable subjects for the Monographs on Theoretical and Applied Genetics are basic and applied aspects of genetic variation; genetic resources; genetic exchange and reproduction; mutagenesis; genotype-environment interaction; gene structure, regulation, action, expression and interaction; chromosomal and extrachromosomal inheritance of economic traits, and genetic models and simulations.

September 1975   The Editors

Preface

Meiotic configurations are looked at from a special point of view in this book: the extraction from them of the maximal amount of quantitative information of genetic interest. Although this requires a certain understanding of their origin and consequences, much of what is known about chromosomes and their formation into the special structures collected under the rather indiscriminate term "configuration", is not considered relevant for this purpose, and simply neglected. Nor is the qualitative significance of meiotic configurations discussed in any detail, for instance their importance as markers of the individual in which they occur. Interpretations of the results obtained, with the approaches discussed, are given only with the purpose of evaluating the analysis or as a preliminary to the further development of an analytical model. The purpose has not been to increase our understanding of chromosome behavior, although at times this may have been a welcome by-product.

The primary objective has been to construct a network of comparatively generally applicable systems for estimating genetic parameters of various natures, but all related to recombination in its widest sense, from relatively simple microscopic observation. Most is based on published reports, some is new. Examples have been taken from various organisms, but no attempt has been made to make the coverage complete. Some excellent publications may well have escaped attention, or they may have been excluded when they did not contribute specifically to the present purpose, or when satisfactory examples had been used from other sources. Some less important work may have been referred to when it contained material or approaches of special interest.

Thanks are due to Miss HENRIËT BOELEMA for her careful typing of the manuscript, to Mr. K. KNOOP for his help with several of the illustrations, and to many colleagues in this department and elsewhere for valuable discussions and for permission to reproduce their photographs.

Wageningen, September 1975   J. SYBENGA

Contents

Chapter 4. The Analysis of Distribution: Centromere Coorientation   . . . 200

Chapter 1. Introduction

Although the reader may be assumed to have a reasonable background knowledge and understanding of the basic meiotic phenomena, it may be useful to recapitulate a few principles. The first Chapter deals primarily with this recapitulation, but it does so from a special angle: that of interpreting meiosis in terms of recombination. It also briefly considers the nature of the information that can be derived from such an interpretation.

1.1 Genetic Variation and Genetic Mechanisms

The units of genetic function (genes, factors, cistrons, operons) appear in different functional forms: *alleles*. One allele can be transformed into another by mutation. Different alleles may differ by one or by more than one consecutive mutational step. They may also be separate and distinct mutant derivatives from a common ancestral form. The expression of most new mutant types is negative in comparison to the established original form, but occasionally mutation results in functional "progress". This may be an inherent positive effect (in the original or in a new environment) or it may result from a favorable interaction with particular alleles of other genes or even, in heterozygotes, with the parental allele. A high rate of mutation destroys the proper functioning of the biochemical system, and all living organisms appear to have mechanisms to prevent genetic damage that might lead to mutations, as well as mechanisms to repair it. There remains a basic mutation rate that on one hand results in a "load" on the capacity of the population to reproduce and maintain itself, but that on the other hand preserves the opportunity for new useful alleles to arise. When the origin of new alleles is infrequent and random, the probability of formation of new complex balanced combinations of alleles is extremely small. This situation is not actually observed: all forms of life have systems by which mutants arisen independently in different individuals can be recombined and the best combination selected. With random recombination, given a relatively large variation, the extraction of well adapted combinations of alleles would still be small but for entirely different reasons than when the new combination depends on new mutation alone. The balance between the immobility of strict conservation and the chaos of random recombination is struck by the *regulated manipulation* of genetic variation by mechanisms of controlled recombination. These mechanisms themselves are under genetic control and subject to selection. Selection, although permitting new systems to evolve, tends to be conservative rather than innovative. Thus, once a system of regulated recombination has been established in a genetic population (either because of its success, or by chance) it may acquire considerable stability, to a great extent because of its own restrictive properties. In this respect it resembles most evolutionary specializations. Only by disruption of the balance, perhaps as a result of chance introduction of new genes by hybridization or mutation, or by changing conditions favoring the selection of exceptional recombinants, is the system of recombination to be changed. Then a change of the entire genetic make-up of the affected population is inevitable, and only gradually can a new balance be established.

Systems of regulation of recombination clearly are of considerable biological importance, and their quantitative description in terms of genetic (recombination) parameters of theoretical and practical interest.

Recombination is (almost) universally associated with reproduction, and is effected in two steps: the combination of two gametes (male and female) into a zygote from which a new individual develops, and the formation of the gametes. Neither the factors affecting the combination of gametes (for instance incompatibility systems preventing some and favoring other gametic combinations) nor the factors affecting the availability of specific gene combinations (selection) are normally considered part of the recombination system proper. The term recombination is usually restricted to phenomena associated with the determination of the genetic composition of the gametes. It means in fact that the gametes combined into a zygote are genetically different from those combined to form the parents, not considering the creation of new or the loss of existing genes.

It appears that the factor of major importance for the regulation of recombination is the organization of the genes in large associations, the *chromosomes*. The properties and the behavior of these chromosomes determine the regulation of recombination. These are determined by factors inherent to the chromosomes themselves and by factors acting on the chromosomes. As far as the latter are genetic factors, they are also situated in the chromosomes, but exert their influence indirectly.

All genes are long, linear molecules of DNA, associated end to end in the chromosomes and, especially in higher organisms, combined with several specific proteins, including basic and acidic proteins, enzymes, RNA and a number of anorganic ions. The necessity of packing large numbers of genes (several thousands) into a small number of chromosomes (usually between 5 and 25) may primarily be a consequence of the impossibility of properly arranging the simultaneous separation of the daughter units after replication of individual genes. The compound structure, in addition, offers opportunities for correlated regulation of replication and transcription. The important *genetic* consequence is that it keeps together specific alleles of different genes, preventing the free and random formation of new allelic combinations except by mutation. Linking genes into chromosomes is the most potent mechanism of restricting recombination, but the linkage is not perfect and may be broken according to certain rules.

Systems of recombination can be indirectly analyzed quantitatively by studying the numerical relationships between different classes of progeny (segregations). In contrast to this approach, this book deals directly with chromosomal structures (*configurations*) visible at meiosis.

Although the biochemistry of recombination has made considerable progress in recent years (cf., for instance, STERN and HOTTA, 1973) it is not yet possible to describe systems of recombination quantitatively on a biochemical basis.

1.2 Meiosis as a Complex of Processes Regulating Recombination

In the simplest case, a (diploid) organism has only two (corresponding, homologous) chromosomes, one received from the father, one from the mother. Both contain a complete set of genes, which in principle differ only at the allelic level. Any gamete formed by this diploid can only have either one or the other of these two chromosomes. At meiosis the diploid nucleus containing the two chromosomes divides, and one chromosome moves to one of the two daughter cells, the second to the other. There is no choice (Fig. 1.1A). If instead of one, there are two chromosomes containing the genes, say A_1 and B_1 from the mother and A_2 and B_2 from the father, the diploid is $A_1A_2B_1B_2$, and then a choice is necessary when at meiosis the chromosomes go to the two daughter cells. The chromosomal combinations can be A_1B_1 and A_2B_2 as in the parents, or, with equal probability, the combinations can be A_1B_2 and A_2B_1 respectively, the two *recombination* types (Fig. 1.1B). There is recombination at the chromosomal level, but the parental types reappear. This system of recombination implies that the recombination types are complementary. With more chromosomes per genome (= a complete set of genes), the probability of recovering the parental types decreases, and the number of different recombination types increases.

Which combination of chromosomes is brought together into one daughter cell is determined by the orientation of the pairs of chromosomes in respect to each other at the stage preceding the separation. There are two prerequisites for this orientation: (1) Association (pairing) of the homologous parental chromosomes into pairs *(bivalents)*. (2) Temporary attachment of the paired chromosomes to permit a stable orientation of the two chromosomes, each onto one of two poles in the cell. In some organisms this attachment is brought about by mere stickiness that keeps closely associated chromosomes effectively together until a biochemical change in the cell releases them. More common is the situation that paired chromosomes split first into two identical longitudinal parallel strands, the two chromatids, which is followed by an exchange of segments between one of the two chromatids of each chromosome (Figs. 1.2; 1.3). This results in a sort of knot *(chiasma)* which keeps the paired chromosomes together as long as the two chromatids do not separate. The chiasma itself is not strong enough. It is kept intact by some form of stickiness between the chromatids. Mere stickiness between the paired chromosomes is not sufficient in chiasmate meiosis without the presence of at least one chiasma. The chiasmata thus have two "functions": they help to keep the chromosomes together until the moment the period of orientation is completed, and, more significantly, represent *exchange of genetic material* between the two homologous chromosomes. The genetic conse-

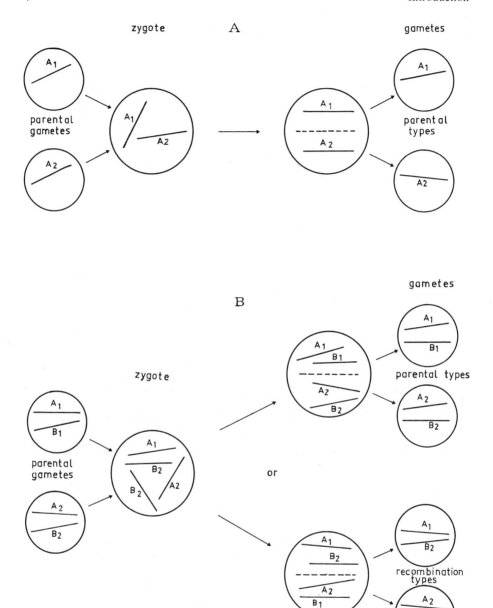

Fig. 1.1 A and B. Drastically simplified diagram of chromosome recombination. (A) With one chromosome per genome, chromosome recombination is excluded. (B) With two chromosomes per genome, besides the two parental combinations, two new (recombination) types can be formed

A   B

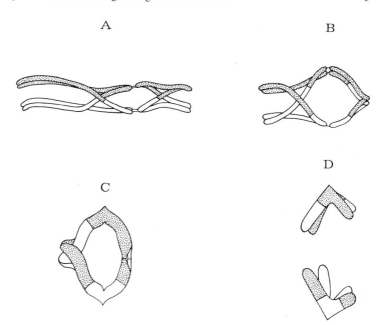

D

C

Fig. 1.2A–D. Two chiasmata in a bivalent of two paired homologous chromosomes. Only one chromatid of each chromosome is involved in each chiasma. The difference in shade reflects allelic differences. The constriction is the centromere, the attachment points of the fibres that pull the chromosomes to the poles. (A) is a diplotene bivalent, (B) the same bivalent at diakinesis: the chiasma has moved from its original position as a result of torsion caused by contraction of the chromosomes. In (C) (metaphase, the stage of orientation of the chromosomes on two poles) this terminalization has proceded even further. At (D) the separation has taken place (anaphase). As a result of crossing-over the chromatids of the two chromosomes are different in respect to allelic constitution. (From SYBENGA, 1972a)

quence of this exchange is called *crossing-over*. The bivalents with their chiasmata can be observed microscopically in favorable material (Figs. 2.1; 2.2).

The two paired chromosomes together have four chromatids. Only two of these experience exchange when one chiasma has formed, the other two remain unchanged (Figs. 1.2; 1.3). The two chromosomes segregating at anaphase thus consist of two chromatids which are genetically different. This first *(reductional)* division, therefore, is normally followed by a second division in which the two chromatids are separated and where each is incorporated into a new daughter nucleus. The result is a tetrad of four cells, all genetically different as far as the chromosomes of the mother and the father carried allelic differences (Fig. 1.3). There is no doubt that exchange is the primary function of the chiasma and that the mechanical function in stabilizing orientation is derived.

There are apparently two systems of recombination at meiosis: *chromosome recombination*, in effect depending on the number of chromosomes, and *within-chromosome* (or between homologue) recombination, in effect depending on the number and location of the chiasmata.

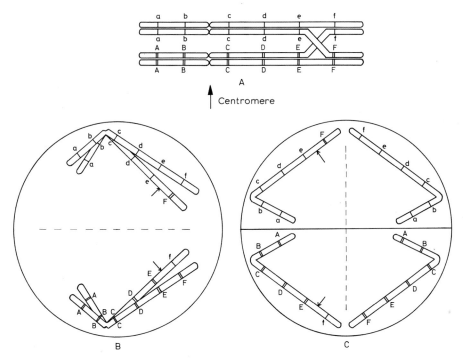

Fig. 1.3 A–C. The fate of the chromatids of a bivalent with one chiasma. Allelic differences are indicated by capital and low cast letters respectively. Note that after completion of the two divisions the two parental types are recovered in addition to the two complementary recombinant types. With more chiasmata the probability of recovering the parental types decreases. The small arrows in (B) and (C) indicate the location of the exchange corresponding with the chiasma. (From SYBENGA, 1972a)

   At meiosis, complete single (haploid) groups of chromosomes are extracted from diploid groups, in order to be included in gametes which will fuse into new diploids. The realization and regulation of recombination, however, are the primary function of meiosis, and the alternation of a diploid with a haploid generation, and consequently the process of reduction at meiosis and fusion at fertilization, are secondary phenomena.

   Meiosis runs as follows (Fig. 1.4): in the nuclear mass of the early primary meiocytes the chromosomes gradually become visible as very thin threads. This first stage is called *leptotene*. Homologous chromosomes in some way manage to find each other in the next stage, *zygotene*, and starting from a few initial points pair along their entire length. Perhaps the homologues usually complete the first stages of mutual attraction at an early condensed stage. Zygotene pairing then is merely the completion of the process. At the following stage, *pachytene*, pairing is complete and the condensation of the bivalents has proceded so far that in organisms with small chromosomes the individual bivalents can be distinguished and sometimes even recognized. Their surface is not smooth but carries the typical *chromomere* pattern of slightly more condensed particles (the chromomeres) alternating with thinner parts. At *diplotene* the homologues

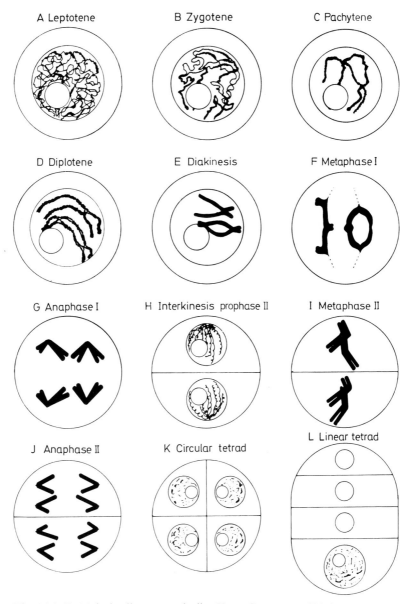

A Leptotene   B Zygotene   C Pachytene

D Diplotene   E Diakinesis   F Metaphase I

G Anaphase I   H Interkinesis prophase II   I Metaphase II

J Anaphase II   K Circular tetrad   L Linear tetrad

Fig. 1.4 A–L. Meiosis, diagrammatically. (From SYBENGA, 1972a)

separate again but are kept together at the chiasmata that have formed presumably at pachytene but that are not visible before diplotene. Each chromosome can now be seen to consist of two chromatids. At this stage the chiasmata are assumed still to have the position at which they originated. With further separation of the chromosomes of the bivalent (first primarily because of torsion, later also as a result of active separation under the influence of the centromere, which is the locus primarily responsible for chromosome movement) the chiasmata tend to change their position. In the beginning they merely move away from each other, making their mutual distances more equal. Later they move primarily towards the chromosome ends (chiasma *terminalization*). Condensation of the chromosomes proceeds and reaches its maximum, usually at diakinesis. This is the last stage of the first meiotic *prophase*, and the nuclear membrane and nucleolus are still intact. They disappear when the spindle is organized, and *prometaphase* starts. At this stage in a number of organisms the bivalents experience considerable stretch (prometaphase stretch) followed by another condensation at *metaphase* proper where the bivalents line up in the equator. The centromeres are well on their way to the poles but the chiasmata prevent them from moving further. The chiasmata at this stage are positioned in the equator, half-way between the poles. This is the stage of *orientation*.

A biochemical change in the cell releases the adhesion between the chromatids and the chiasmata slip off: the chromosomes are free to move to the poles *(anaphase)*. At anaphase, two chromatids can clearly be seen in each chromosome. After anaphase, telophase and interphase (here often called *interkinesis*) follow. Interkinesis is usually quite short since chromosome replication is not required, even undesired. Sometimes first telophase proceeds directly into second prophase. The whole second division closely resembles a mitotic division, except for an often somewhat less clean metaphase: the chromatids are usually not closely aligned but already somewhat separated.

For holokinetic chromosomes, which in stead of a single centromere, have their kinetic activity distributed evenly about the chromosome, the process is slightly different, especially as regards orientation. Basically, however, it is the same.

The result of meiosis in principle is a tetrad of four cells. In plants this is indeed formed, usually circular in the case of pollen mother cells, where each daughter cell produces a pollen grain. In embryosac mother cells the tetrad is linear and usually one of the two terminal cells develops into an embryosac, while the others degenerate.

In animals, in spermatocytes the products of the first division separate and the second division proceeds in the two daughter cells separately: there is no tetrad, but all four products may be functional. In the oocytes one of the two daughter cells of the first division degenerates into the first polar body and only the other one proceeds with the second division. Here again one of the products degenerates (second polar body) and only one nucleus remains to form the egg.

1.3 The Three Central Elements of Meiosis: Chromosome Pairing, Crossing-over, Distribution

Some of the meiotic stages can vary quantitatively without genetic consequences. Anaphase, for instance, may be long or short, may have a wide or a normal spindle, but all it does is to transport the chromosomes to the poles. Whatever variation there is either does not affect the end result or it results in an abnormality.

First meiotic metaphase, on the other hand, determines *orientation*, i.e. determines the poles to which the chromosomes are distributed at anaphase, and this is a meiotic element essential for the genetic composition of the gametes. It may seem that in respect to orientation different bivalents are independent and that distribution must therefore follow a system based on randomness, not influenced by internal or external factors and that distribution thus must obey statistical rules of a rigidity similar to those of simple transport. There appear to be many exceptions, however, where orientation is systematically affected in specific ways. This variation in orientation at first metaphase has a component due to chromosome structure, and in addition is under general genetic control. Distribution at second anaphase may also at times have important consequences, but there seldom is a deviation from randomness. Orientation-dependent distribution thus is a central element in meiosis, exposed to variation with genetic consequences, and therefore of central importance for the regulation of recombination.

Crossing-over, cytologically recognized by the presence of chiasmata, is also variable, both in respect to *frequency* and to *localization* of the chiasmata. Variation in crossing-over affects the recombination system and in more extreme cases can affect orientation not merely when chiasmata fail in critical segments, but especially in cases of complex chromosome associations by affecting size and shape of the associations. This again is important for distribution. For two reasons, therefore, the systems of crossing-over are of importance for the genetic composition of the gametes.

Chromosome pairing seems at a first glance to be a process that affects the proper functioning of meiosis only in a qualitative way: when there is no pairing, there is no basis for crossing-over and for distribution. There are two ways, however, by which variation in chromosome pairing may have quantitative genetic consequences: (1) Incomplete pairing, barely enough for the minimum number of chiasmata required for proper orientation, restricts recombination. The extent of pairing may be under genetic control and may be influenced by environmental factors. (2) In competitive situations, where more than two homologues are present and compete for pairing, variation in the pairing system can have a profound effect on the choice of pairing partners, and consequently not only on recombination but also on distribution. Variation in pairing, therefore, has genetic consequences and must be considered the third factor of central importance for the genetic composition of the gametes.

The chronological order is: pairing—crossing-over—distribution, and in this order each influences the other: pairing sets a limit to crossing-over, and crossing-over to orientation and thus distribution. Except for this hierarchic relation all three can vary independently. They are under the control of a (genetic) system of regulation which itself is to a great extent determined by the particular restrictions it imposes on these processes.

1.4 The Order of Analysis: Crossing-over—Chromosome Pairing—
Distribution

Although chronologically the first of these three processes is chromosome pairing, it can almost never be analyzed quantitatively at the moment it takes place. Occasionally, when meiotic pairing is preceded by somatic associations that predetermine meiotic association, analysis of somatic pairing makes it possible to predict how meiotic pairing may be expected to proceed. Electron micrographs may occasionally show chromosomes in the process of pairing, but they are not favorable for quantitative observation. Pachytene is accessible in a number of organisms with few chromosomes, but non-homologous association, not comparable with effective pairing leading to genetic exchange, often disturbs the analysis. Usually it is inevitable to resort to later stages, where the association of chromosomes has become dependent on the presence of chiasmata. When there are chiasmata, there must have been pairing, but when there are no chiasmata, not much can be said about pairing. It is useful, therefore, first to consider the analysis of crossing-over (chiasmata) and then to analyze pairing, making use of the information available on chiasmate association. This is the reason that in the following chapters the analysis of crossing-over preceeds that of pairing even though this order does have certain disadvantages.

1.5 The Nature and Quality of the Information Derived from Meiotic
Configurations

Meiotic configurations are the apparent forms in which chromosomes or, more significantly, combinations of chromosomes are present at meiosis. This is a wide interpretation and includes all stages at which the chromosomes can be distinguished. As a result of variation in pairing, chiasma formation and orientation, the same chromosomes can form different configurations at the same stage in different meiocytes of the same individual. The relative frequencies of these different configurations can be used for describing quantitatively the characteristics of the particular systems of pairing, chiasma formation, and orientation. The configuration, in this context is not used as a *marker*, i.e. the information derived from the configuration in respect to the genetic or chromosomal constitution of the individual in which it occurs is not relevant. This can perhaps be made clear by the following example: when in a population a proportion of the individuals shows a quadrivalent in a certain percentage of its meiocytes, whereas other individuals of this population do not show a quadrivalent in any of their cells, it is reasonable to assume that the individuals with a quadrivalent carry a reciprocal translocation (interchange) in heterozygous condition. The occurrence of the quadrivalent marks the individual as an interchange heterozygote. This is merely qualitative information and it can be used for instance in population studies. It has no relevance in respect to chromosome pairing, crossing-over or segregation in the individual on which the observations are made.

In one important aspect the genetic information derived from observations on meiotic configurations differs from that obtained in classical genetic analysis. In classical genetics the segregational relations of *genes* are studied. In the present approach genes are not recognized, but the behavior of more or less distinct and specified segments of the genome is analyzed. These segments carry genes, and therefore the behavior of these segments is genetically relevant. What this means in terms of specific genes, however, cannot be determined.

The quality of the information derived from meiotic configuration frequencies depends on the accessibility of the chromosomes to observation, on the number of observations available, on the opportunities for constructing special cytological stocks and on the soundness of the theoretical models used. There is a great variation in accessibility. In some organisms, early prophase can be analyzed in detail and then not only pairing can be studied, but the frequency of chiasmata can be determined exactly, and their location recorded. Even the chromatids involved in each consecutive chiasma can be recognized. On the other extreme of the scale are the organisms where nothing more can be observed than whether or not chromosomes are associated at all, without knowing how many chiasmata are present, nor which arm, nor even which chromosome is involved. Although in such case the quality of the information is very limited, there are some indirect methods of estimating, for instance, the recombination capacity of the organism. In the first place the number of chromosomes itself: it determines the extent of chromosome recombination. The more chromosomes, the more recombination in the genome. In the second place, the fact that two chromosomes are associated means that at least one chiasma must be present. The ratio between chromosomes (or chromosome arms) with chiasmata and those without gives an impression of the level of the within-chromosome recombination. Although at first sight the information thus obtained may seem quite limited, there are methods for maximizing it (Section 2.1.2).

When special chromosomal constructions are available, they can greatly increase the resolution of the analysis, either by *marking* specific chromosome segments or by creating conditions that make the recognition of *relevant events* possible.

Thus, in respect to recombination, roughly three levels of resolution (with transitions) can be distinguished:

1. Direct and exact determination of number and location of chiasmata and of the chromatids involved, in specified chromosomes (2.1.1).
2. Inferential but rather exact determination of the frequency of specific (marked) chromosome segments having at least one chiasma (2.2; 2.3).
3. Indirect estimation of recombination from bivalent number and, to a more limited extent, from associations of chromosomes in bivalents in unmarked material.

Similar differences in resolution can be distinguished in respect to pairing and orientation.

At all levels of resolution doubt about the correct interpretation, and the possibility that essential changes have occurred between the event and the observation, decrease the reliability of the estimates. For instance, when observing chiasmata in bivalents of the usually very favorable locusts *Locusta migratoria*, or *Schistocerca gregaria*, there are often some bivalents in which it is not certain whether or not a point where the chromosomes appear to touch is a true chiasma or merely a twist in the bivalent. At later meiotic stages, especially in deviant types like haploids and interspecific hybrids with high degrees of asynapsis (non-pairing of chromosomes) it is often not certain whether certain associations are true chiasmata or not. Or, when recording the location of a chiasma in interference studies, there is no complete certainty that the chiasma has not shifted its location. Or, at late metaphase some chiasmata may have slipped off and as a result, a ring bivalent of a pair of metacentric chromosomes (chiasmata in both arms) may have changed into an open bivalent (chiasmata in one arm only, Fig. 1.5). The same may happen in more complex configurations

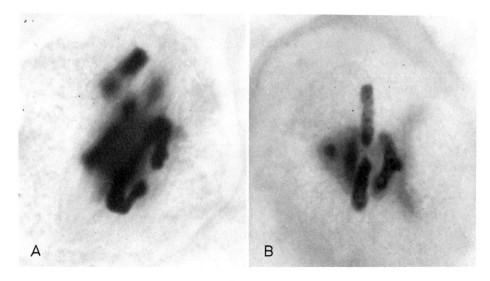

Fig. 1.5A and B. Microtome sections of pollen mother cells at first meiotic metaphase in *Secale cereale*. (A) A bivalent with a chiasma in both ends of a metacentric chromosome loses one chiasma, although the cell as a whole has not yet proceded to anaphase. Slightly later such a bivalent would have been recorded as a bivalent with one chiasma. (B) A similar loss of a chiasma from a bivalent which originally had only one: it is converted into two univalents. The later in metaphase, the more frequently this occurs. In squash preparations such spurious univalents may become separated and scored as a pair of chromosomes without chiasmata. (From SYBENGA, 1958)

(SYBENGA, 1968; DE BOER, 1975, and many others). Finally, some complex configurations, when seen from a critical angle may strikingly resemble an entirely different configuration, from which it may be readily distinguished under more favorable conditions of observation.

When meiotic configurations are used as a source of quantitative information on genetic mechanisms these two sources of error (misinterpretation and changes during the interval between event and observation) should always receive thorough attention.

Chapter 2. The Analysis of Crossing-over

The analysis of crossing-over preceeds that of chromosome pairing for reasons discussed in Chapter I. In Chapter II direct observations on crossing-over (number and place of chiasmata) are considered as well as more indirect observations (frequencies of chiasmate association, derived from the relative frequencies of specific configurations). Only few organisms lend themselves to a satisfactory analysis of meiotic stages at which chiasmata can be individually observed at the locations where they may be assumed to have been formed. Much attention is therefore given to situations in which the original number and position of chiasmata can *not* be observed. These require a special analytic and statistical approach. The amount of information thus extracted from normal bivalents is limited. Therefore, ample use is made of special chromosome constructions which result in larger configurations. These serve a dual purpose: They mark the chromosome or chromosome segment of interest in cases where recognition is not otherwise possible, by generating specific configurations. Here the frequency of chiasmate association of specific segments can be estimated by comparing the frequencies of different configurations. Secondly, they create special conditions in which the effect of deviant chromosome behavior on recombination can be studied. It is of importance that at least some segments must remain virtually unaffected by the abnormal condition, if effective use of the marking function of the construction is to be made. Whenever possible, the analysis has included the interrelations between the different segments of both simple and complex configurations (interference). Inspection of the Table of Contents makes it clear that a great variety of such special constructions have been discussed. It may be noted however, that neither in theory nor in application, have all been worked out in equal detail.

2.1 Unmarked Chromosomes

2.1.1 Crossing-over Estimated Directly from Chiasma Frequencies

The presence of a chiasma marks the event of genetic crossing-over. Counting chiasmata in a distinct segment of a chromosome is equivalent to measuring crossing-over in that segment. Whereas in each chiasma only two of the four chromatids of a bivalent are involved in crossing-over, in genetic experiments all four are recovered individually in the progeny. Of these, only two are cross-over chromatids and two are non-cross-over chromatids. One chiasma then corresponds with crossing-over in only half of the progeny (Figs. 1.2; 1.3), and to convert chiasma frequencies into crossing-over frequencies, they must be divided by two.

As stated in 1.5, there are important differences between estimating crossing-over in genetic experiments using recombination between marker genes, and in cytological analyses based on counting chiasmata. These differences relate to the analysis as well as to the results.

For estimating a single recombination frequency from a genetic classification, a population of observations (individuals) is required, and no information is obtained in respect to distribution and variation of recombination. When studying chiasmata, each observation (cell, chromosome, chromosome segment) is a replication and yields an individual estimate (0, 1, 2, 3 etc. chiasmata) and a population of cells, chromosomes, or segments, yields information not only in respect to the average crossing-over frequency, but also in respect to its variation, or to the distribution of cross-overs. In favorable material the actual location of the cross-overs in a chromosome segment can be determined. Very detailed genetic experiments involving more than two loci with short distances between the loci and consequently hardly ever more than one cross-over in each segment can also give quite detailed information on the distribution of cross-overs, but fundamentally the approach remains indirect. When the purpose of the analysis is to obtain information on recombination between the specific markers involved, of course the genetic analysis is superior. Very often, however, the level of recombination and the distribution of cross-overs in particular chromosomes or in specific chromosome segments are of more general genetic importance. Then observations on chiasmata are superior. Few organisms are known in which both genetic and chiasma analysis can be performed in great detail. Some are especially favorable for genetic studies (Drosophila, maize, barley to name a few), others are particularly favorable for chiasma studies. In maize, which is well studied genetically, with some care chiasma counts can be carried out with reasonable accuracy at diakinesis and metaphase. Already in 1934, DARLINGTON counted on an average 27.05 chiasmata per cell, which corresponds with 13.53 cross-overs, or 1,353 units of map length. At that time the total genetically estimated map length was only 618 units. At present it can be estimated at approximately 1,100, and when more genes become known, this may even increase slightly. The discrepancy between 1,100 and 1,353 has several causes: (1) When marker genes are not very close together, double cross-overs will cause their distance to be underestimated. Sufficiently closely linked markers are not available for all segments to exclude double crossing-over. (2) The markers used do not include the terminal segments of all chromosomes. (3) Chiasma frequencies were determined in pollen mother cells, whereas crossing-over was determined in both female and male parents. Although recombination has been shown to be higher in male recombination in at least one specific chromosome segment in maize, the general rule in plants (as well as in animals) for the genome as a whole, is that female recombination is higher (cf. 2.1.1.5).

A more specific comparison between the frequencies of chiasmata and of crossing-over was reported by FU and SEARS (1973). In a monotelocentric (one chromosome of a pair replaced by the telocentric corresponding with one arm, cf. 2.2.1) of wheat, the frequency of chiasmata in the heteromorphic bivalent consisting of a normal and a telocentric chromosome could be scored at diakinesis. Crossing-over between the centromere and a dominant terminal marker could be estimated in a genetic analysis, involving the same telocentric. The correspondence was good: in 2,681 PMCs at diakinesis, 95.7% had a chiasma in a heteromorphic 4A bivalent. In a testcross progeny of 500 plants, on the basis of this frequency 239.3 were expected to be genetic cross-overs, the remaining 260.7 being non-cross-overs. The observed numbers were 227 and 273 respectively. In a heteromorphic bivalent of chromosome 6A (6,615 PMCs), 97.2% chiasmata were observed, corresponding with 446.3 cross-overs in the 918 testcross progeny plants grown. Observed were 424. Metaphase chiasma counts were very variable and always much lower than diakinesis counts. With elevated temperatures, chiasma frequency and crossing-over decreased. Even at diakinesis, however, the scores were significantly lower than corresponded with crossing-over. Apparently, in heteromorphic bivalents of wheat, chiasmata tend to get lost before metaphase under normal conditions, and even before diakinesis under unfavorable conditions.

In the great majority of organisms few or no genetic markers are available but in most at least some observations on the chiasmata can be made. In principle, all meiotic stages from early diplotene to first metaphase are available for analysis, but there is a great variation in respect to actual accessibility. In only few organisms is diplotene accessible (several salamanders and locusts (Fig. 2.1) for instance, and, in plants, some Allium and Lilium species). Usually the nucleus at this stage is still an entangled ball, or when the bivalents can be separated the chromatids are not clear and a chiasma cannot be distinguished from a mere twist in the bivalent. Then the more contracted stages (diakinesis, metaphase I) may be better, but the compact shape of the configuration often makes it difficult to distinguish different chiasmata, and then all that can be concluded is that either there are some (one or more) chiasmata in the segment studied, or none at all.

One point of considerable interest is the fate of the chiasma between formation and observation. It is generally assumed that chiasmata move towards the chromosome ends (terminalization, DARLINGTON, 1965), but the process is not simple. It is very probable that at early diplotene, when the chromatids around the chiasmata are not closely apposed, a chiasma can move rather freely from its point of origin. As the forces of torsion of the condensing bivalent (causing this movement) are not yet strong, the movement will be rather limited, but may not be neglected. Since these forces are stronger in the small loops between closely neighboring chiasmata than in the large loops between chiasmata further apart, the short loops will tend to increase in size at the expense of the longer loops and the chiasmata will become more equidistantly positioned. When with proceeding condensation the forces of torsion increase, so does the adhesion between associated chromatids. These are sister chromatids in the sections between chiasmata where no chiasma movement has taken place, but non-sister chromatids where a chiasma has passed. Since the relation between increase in torsion and increase in adhesion between chromatids is not necessarily the same for different species, nor for different individuals of the same species

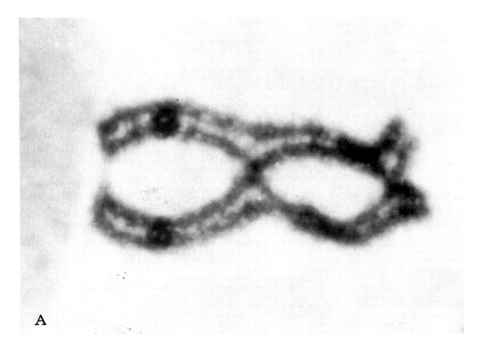

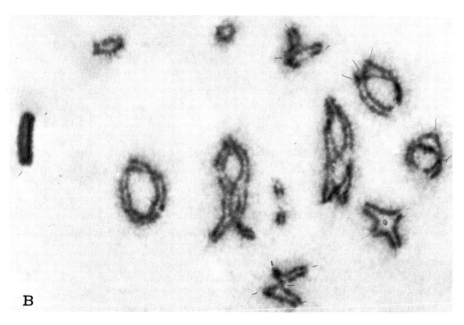

Fig. 2.1. (A) A diplotene bivalent in a spermatocyte of the salamander *Oedipina poelzi.* Darkly stained area marks the centromere. (Courtesy J. KEZER). (B) Bivalents (and X-univalent) in a spermatocyte of the locust *Schistocerca gregaria*. Chromosomes acrocentric, centromere not recognizable. (Courtesy S. A. HENDERSON). In both examples the number and location of, and even the course of the chromatids in, the chiasmata can be observed

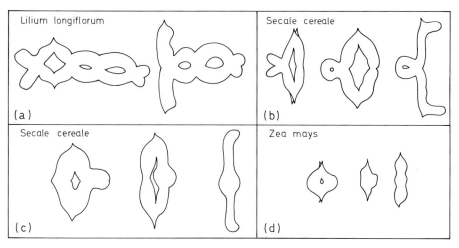

Fig. 2.2a–d. Examples of different types of metaphase bivalents. In plants and animals very similar shapes are observed. In (a) the number of the chiasmata can readily be determined; their location has probably shifted only slightly since diplotene. In (b) the number can still be determined with reasonable accuracy but there may be doubt about the location. In the same species under different genetic or environmental conditions, or even in the same cells, distinction between neighboring chiasmata may become uncertain (c). In (d), although occasionally different chiasmata in the same arm can be distinguished, in general it is not possible to say more than that an arm either is associated (or "bound") by chiasmata or not

nor even for different bivalents in an individual, it is rather difficult to estimate the amount of movement when comparison between early and late stages is not possible. HEARNE and HUSKINS (1935) found considerable movement of chiasmata between early and late diplotene in the locust *Melanoplus femur-rubrum*, but only little movement at later stages. In another locust *Schistocerca gregaria* (Fig. 2.1) HENDERSON (1963) and FOX (1973) considered movement at diplotene of insufficient importance to be taken into account in their chiasma localization studies.

At the onset of prometaphase when the spindle is organized and the centromeres start separating, a second force is added to the one originating from torsion. While torsion in the bivalent resulting from chromosome contraction moves the chiasmata away from each other and the terminal chiasmata towards the ends, the centrometric forces move the chiasmata away from the centromere and towards the end of the chromosome arms.

There is no agreement on the extent of movement of chiasmata at more condensed stages. Clearly, in large bivalents with numerous chiasmata, such as in several lilies (Fig. 2.2), the broad bean *Vicia faba*, and some salamanders, even at metaphase the chiasmata appear to have experienced little movement. They may have shifted their position slightly at early diplotene, resulting in a more even distribution of their relative positions. Whether or not it is at all possible that adjacent reciprocal chiasmata cancel each other when they are close enough, is not known. From what can be seen of diplotene chiasmata

it is probable that even the nearest two chiasmata are far enough apart to
move away from each other instead of fusing (Fig. 2.5e). It is probable, therefore,
that no significant reduction in chiasma frequency occurs in such large bivalents
between diplotene and metaphase. Observational difficulties remain, but are
mainly due to the impossibility of distinguishing between twists in the bivalents
and true chiasmata. The few cases in which differences between stages have
been reported in such material can be explained convincingly by misclassification
at diplotene, where twists in the bivalents were counted as chiasmata (LEVAN,
1936).

More serious difficulties are encountered in metaphase bivalents with relatively
few and predominantly distally (away from the centromere) positioned and terminalizing chiasmata. They may sometimes be distinguished as separate chiasmata
(Fig. 2.2A, B), but often it is quite difficult to decide whether one or two
are present (Fig. 2.2C, D). In some species chiasmata are always completely
terminalized, and there is no way of knowing how many there have been when
earlier stages are not accessible. It is often assumed that there have never been
more than one in an arm of such bivalents, but it is quite possible that two
have "fused" into one terminal attachment point. For a stable attachment close
approximation of chromosomes by means of chiasmata is merely a preliminary
to a biochemically regulated sticking-together. When there is uncertainty about
chiasmata having fused or not there is no use in counting chiasmata. It is
more sensible simply to consider the frequency of metaphase *association* and
to use this information for estimating crossing-over (Section 2.1.2).

2.1.1.1 Chiasma Frequency Estimates

There are numerous observations on diplotene chiasma frequencies, mainly in
locusts. Table 2.1 for example gives the results of an analysis by HEARNE and
HUSKINS (1935) of *Melanoplus femur-rubrum*. Their diakinesis and metaphase
data are practically identical to the diplotene results. The chromosomes are
acrocentric: there is only a single arm. Comparable data are given by HENDERSON

Table 2.1. Average chiasma frequencies per bivalent for the 4 long, 4 medium and 3 short
bivalents of *Melanoplus femur-rubrum* at different stages of meiosis. (Calculated from HEARNE
and HUSKINS, 1935)

Stage	Bivalents			
	Long	Medium	Short	No. of cells
Early diplotene	1.766	1.047	1.000	16
Mid diplotene	1.714	1.036	1.000	21
Late diplotene	1.646	1.021	1.000	12
Diakinesis	1.731	1.019	1.000	13
Metaphase I	1.655	1.036	1.000	21

(1963) and FOX (1973) for *Schistocerca gregaria*, (Fig. 2.1 B) without distinguishing different diplotene stages, but treating all 11 bivalents separately (Table 2.2). In both objects, the short bivalents have consistently one and only one chiasma. The medium and long bivalents have more chiasmata in Schistocerca than in Melanoplus.

Table 2.2. Average chiasma frequencies and variances for the 11 bivalents separately at diplotene in *Schistocerca gregaria*. Ten males, 25 cells each (HENDERSON, 1963). L = long; M = medium; S = short

Bivalent	L_1	L_2	L_3	M_4	M_5	M_6	M_7	M_8	S_9	S_{10}	S_{11}	per cell
Average	3.04	2.60	2.31	2.05	1.92	1.85	1.69	1.34	1.00	1.00	1.00	19.81
Variance	0.002	0.003	0.004	0.003	0.003	0.002	0.001	0.003	0.000	0.000	0.000	0.109

Vicia faba has large bivalents in which numerous chiasmata can be observed at first metaphase. Several reports of metaphase chiasma frequencies are available. According to SAX (1935), the five short acrocentric chromosomes in 60 pollen mother cells had on an average 2.63 chiasmata. The single long metacentric chromosome, of which the two arms could not be distinguished, had 6.75 chiasmata. This is relatively low in comparison to the data of MAEDA (1930) who found 3.5 in the short and 8.1 in the long chromosome, but comparable to those of ROWLANDS (1958), Table 2.4b. The difference may be due to varietal and to environmental differences. Such high frequencies are not uncommon in plants with long chromosomes, as often found for instance among the Liliaceae.

Numerous plants have been studied in which the chiasmata are much fewer in number. Then they are often localized, proximally or distally. An example of the latter is rye; the data presented here have been collected by REES (1955) on inbred and hybrid material (Figs. 2.2B, C) and MATHER and LAMM (1935) on population rye (Table 2.9). There may be some doubt about the number of chiasmata in some of the bivalents, as occasionally it is difficult to decide between a single chiasmata and two closely apposed chiasmata. Rye, therefore, will be encountered again in Section 2.1.2, where not the chiasmata as such, but merely association by chiasmata is considered.

For a meaningful genetic interpretation of chiasma frequencies it is useful to be able to recognize individual chromosomes. This is not always possible. The individual rye bivalents for instance cannot be distinguished in meiosis. In *Vicia faba* the single long bivalent is easily recognized, but the five short, acrocentric bivalents are almost identical. In Melanoplus classification is possible in three groups, and in Schistocerca all bivalents can be recognized, although not consistently as far as the small ones are concerned (FOX, 1973).

The smallest recognizable segment in a normal bivalent is the *arm*, but again the extent to which recognition is possible is very variable. In pronounced acrocentrics, of course, there is only one arm (for instance Schistocerca, Melanoplus and the short bivalents of *Vicia faba*). In metacentric chromosomes, although there are two arms, they are frequently too similar to be distinguished (the long bivalent of *Vicia faba*). In subacrocentrics this may be possible (Fig.

2.2A), but this is not consistently the case with the two sub-acrocentric rye bivalents even though the mitotic arm ratio may approach 2. Sometimes, detailed observations are possible: in BENNETT'S (1938) *Fritillaria chitralensis* in a submeta-centric bivalent, the long arm on an average had 3.70 chiasmata (in 50 cells), the short arm 1.78. For the B chromosome in rye, SYBENGA and DE VRIES (1972) reported 0.86 chiasmata for the long arm and 0.12 for the short arm. More details were reported by ELLIOTT (1958) on *Endymion non-scriptus*, where in most of the chromosomes the long and short arms differed markedly morpho-logically (Table 2.3).

Table 2.3. Mean chiasma frequencies and variances for the whole bivalents and for their separate arms in *Endymion non-scriptus*, plant *MW 6*, together with mitotic lengths of the chromosomes as a percentage of the length of the haploid set (ELLIOTT, 1958)

Chromosome (arm)	Length	Mean Xa frequency	Variance
H	7.3	1.65	0.2333
G	9.0	1.68	0.2763
F	10.8	2.15	0.2846
E	13.1	2.13	0.1122
BCD	13.9	2.64	0.5512
A	18.0	2.88	0.4711
G short	0.9	0	0
BCD short	2.0	0.37	0.2342
H short	2.4	0.65	0.2333
F short	3.2	0.93	0.0712
H long	4.9	1.00	0
A short	5.2	0.93	0.0712
E short	6.0	1.00	0
E long	7.1	1.13	0.1122
F long	7.6	1.23	0.1788
G long	8.1	1.68	0.2763
BCD long	11.9	2.28	0.4532
A long	12.8	1.95	0.3051

Chiasma frequencies will be encountered again in the discussions of interfer-ence, in those on the relation between chromosome length and chiasma frequency, and in those on modifying factors.

2.1.1.2 Interference within Chromosomes

There are two main types: *chromatid interference* and *chiasma interference*. In the former, chromatids are not free to participate at random in two neighboring chiasmata. Fig. 2.3 gives the four types of combinations of two chiasmata. In one type, two chromatids are involved in both chiasmata, the other two chromatids in none. The cross-overs cancel each other and the result is total

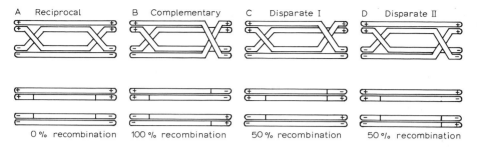

Fig. 2.3 A–D. The four combinations of two chiasmata

absence of recombination between loci outside the two chiasmata (the reciprocal type: two strand double). In the second, two chromatids are involved in one chiasma, and the other two in the other (the complementary type: four strand double; all chromatids are cross-over chromatids). In the third and fourth types, one of the two chromatids involved in one chiasma is also involved in the second, but the other is not (the disparate types: three strand doubles, two chromatids are cross-over chromatids, the other two are non-cross-overs). When all four types are equally frequent, 50% of the chromatids are cross-over chromatids and 50% are non-cross-over chromatids. With common chromatid interference there is an excess of the four-strand doubles and, somewhat less, of three-strand doubles, and this results in recombination percentages of over 50%. These have been reported for fungi, but reports are scare and inconsistent for higher organisms. It appears impossible, in a cytological preparation of stages later than early diplotene, to distinguish between the two-strand and the four-strand doubles (both are "compensating" chiasmata). Although in a "good" squash preparation figures permitting chromatid identification as in Fig. 2.3 can be obtained (Fig. 2.1), and superficially a distinction between two-strand doubles and four-strand doubles would seem possible, such figures are in fact artifacts. In reality, in the three-dimensional bivalent before squashing, a segment on one side of a chiasma tends to stand perpendicular to the segment on the other side of the chiasma from late diplotene on (cf. DOUGLAS, 1970), and this makes it impossible to decide which are the cross-over chromatids in each chiasma, and which the non-cross-over chromatids. This is particularly clear when the chiasma opens out (Fig. 2.4 B). At early diplotene, there are more opportunities, theoretically, but the analysis is usually rather unreliable, and no reports are known to this author. The three-strand doubles (non-compensating), however, can be distinguished from the other two, and an excess (or shortage) of these above 50% indicates some form of chromatid interference, even though its character cannot be decided upon. In the preparation, the non-compensating combinations of chiasmata are characterized by an odd number of chromatid twists more in one chromosome of the inter-chiasma segment than in the other chromosome. The compensating (two- and four-strand double) combinations have either the same number of twists, or differ by an even number (Fig. 2.5).

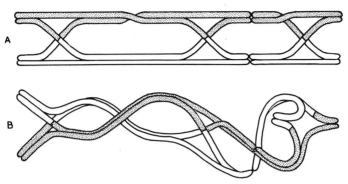

Fig. 2.4. (A) Diagram of a bivalent with three chiasmata; the one in the middle is disparate in relation to both outside chiasmata. (B) The diplotene situation. The end segments tend to stand perpendicular to the loop at the inside of the distal chiasma. At left, after squashing the cross-over chromatids are still in the middle, but if the top end had been forced to the bottom, the crossing chromatids would have been the *non-cross-over* chromatids. At right, the chiasma has opened up, and cross-over and non-cross-over chromatids can by no means be distinguished. (Sketched after EMERSON, 1969)

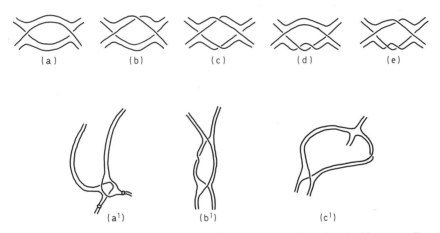

Fig. 2.5a–e. Compensating (a, c, d) and non-compensating (b, e) chiasmata. Examples of the corresponding combinations (a^1, b^1, c^1) in bivalents of *Melanoplus femur-rubrum*. (After HEARNE and HUSKINS, 1935)

In *Melanoplus femur-rubrum* HEARNE and HUSKINS (1935) found 71 compensating pairs of chiasmata and 35 non-compensating. The deviation from 1:1 is highly significant. In *Trillium erectum* HUSKINS and NEWCOMBE (1941) found 279 compensating and 162 non-compensating combinations, again a highly significant difference, and in the same direction. The most simple type (without any chromosome twists between the chiasmata, one of the compensating types, Fig. 2.5) was the most frequent. It appeared that the distance between chiasmata was the shortest for this type, and increased with an increasing number of twists. This has as a result that with increasing distance the excess of compensating combinations tended to decrease. This agrees with the observation that with decreasing chiasma frequency (for instance accompanying increasing chiasma interference, and resulting in larger distances between chiasmata) the compensating types became lower in frequency, i. e. chromatid interference was reduced. This suggests that, in spite of the biochemical and structural complexities of chiasma formation (synaptonemal complexes; several meiotic periods which are critical for chiasma formation; numerous enzymes involved) mechanical factors are of at least some importance.

Chiasma interference, corresponding with the common positive genetic interference, is universal. It can be cytologically detected and quantitatively analyzed in several ways. A deviation from random location of chiasmata may often be assumed to be due to chiasma interference. HENDERSON (1963) in his careful analysis of chiasma position in *Schistocerca gregaria* assumed that movement of chiasmata at diplotene is negligible, and that the striking tendency towards equidistance between chiasmata is a result of interference rather than of movement. He mapped the location of chiasmata in all 11 acrocentric bivalents in 100 cells of adult males. It appeared that when more than one chiasma was formed in a bivalent, the 5% of length nearest the distal end would almost invariably contain a chiasma. The remaining chiasmata were distributed at rather equal distances about the bivalent. The long bivalents had a mode of three chiasmata, the medium bivalents of two and the short bivalents invariably had only one chiasma. These data confirmed MATHER'S (1940) hypothesis of sequential formation of chiasmata from an initial point. Whereas MATHER (1937, 1940) assumed that the initial point was the centromere, HENDERSON concluded that in most cases it was the chromosome end. For bivalents with an average of more than one chiasma the differential distance (d) between the starting point (here the chromosome end) and the first chiasma approached 0. The interference distance (i) between subsequent chiasmata (corresponding to $\frac{1}{2} \times$ the intercept length between subsequent cross-overs) was found to be 7.3 μ of diplotene bivalent length. This is the average distance between chiasmata, which appeared to be rather constant also for different bivalents. There was appreciable variation in the actual distance between chiasmata, but this was for a great part related to differences in number of chiasmata. The more chiasmata are formed in a bivalent, the closer they are together. When the distance between the first and second chiasma happened to be small in the medium bivalents, there was place for a third chiasma. For similar reasons the long bivalents could occasionally have four. Although as a consequence of this variation there was no strict restriction of chiasma formation to specific distances from the initiation point, a definite wave-pattern of distribution could be observed (Fig. 2.6). Besides variation in distance between chiasmata, there was another factor which disturbed the regularity of the system: in 30–40% of the bivalents chiasma formation could be assumed to start at both centromeric and distal ends of the bivalent, causing an irregular interference pattern with relatively high densities at both ends. Occassionally, diakinesis configurations permit an acceptable determination of chiasma location. HULTÉN (1974) carried out an analysis of chiasma distribution at diakinesis in the normal human male, comparable to that reported by HENDERSON (1963) on Schistocerca. The human complement comprises metacentric as well as acrocentric chromosomes, and many intermediate shapes.

When there is reason for considerable doubt about the correspondence of position at the moment of observation and of formation, chiasma interference can be studied in a more indirect way. The first analysis of this kind is by HALDANE (1931), of data published by MAEDA (1930). HALDANE compared the distribution of the numbers of chiasmata per bivalent at metaphase in the long metacentric bivalent of *Vicia faba* with a random (Poisson) distribution (Table 2.4). The classes of high numbers, and those of low numbers were much

Table 2.4(A). The distribution of chiasmata at metaphase in the long metacentric chromosome of *Vicia faba*, compared with a Poisson distribution (HALDANE, 1931; observations by MAEDA, 1930)

No. chiasmata	<3	3	4	5	6	7	8	9	10	11	12	>12
No. of bivalents	0	1	3	7	13	33	39	41	28	8	8	0
Random (Poisson)	1.7	4.0	8.5	14.2	19.9	23.9	25.0	23.3	19.6	15.0	10.5	15.3

Table 2.4(B). Comparison of mean chiasma frequency per bivalent and variance in four plants of *Vicia faba* (ROWLANDS, 1958)

Plant	*m* chromosomes		*M* chromosomes	
	mean	variance	mean	variance
J 0	2.77	0.2548	6.15	0.8275
J 2	2.69	0.3680	7.05	1.4475
J 4	2.68	0.7870	5.65	1.0275
J 8	2.81	0.4940	6.05	1.1475
Mean	2.74		6.23	

less frequent than expected. This was interpreted to mean that the chiasmata at the moment of formation had tended to stay away from each other, which resulted in the impossibility of high numbers per bivalent being realized. The absence of the expected low numbers can be explained as follows. The average observed, as a result of the absence of high numbers, is necessarily lower than it would have been had there not been interference. When calculating the "expected" frequencies on the basis of this low average, the frequency of bivalents with very low numbers of chiasmata is over-estimated. The significance of the difference between observed and random distributions is most conveniently estimated in a X^2 test. There is no certainty, however, that *all* of the difference is due to interference.

Instead of comparing entire distributions, interference can equally well be detected by comparing the observed variance with the expected variance. The latter, in a Poisson distribution is equal to the mean. The ratio between observed

(V_o) and expected (V_e) variance equals the *coincidence*: $\left(\dfrac{\text{Variance}}{\text{Mean}}\right) = \dfrac{V_o}{V_e} = c$, and

$1-c$ is the index of interference. The test of significance is the same as that

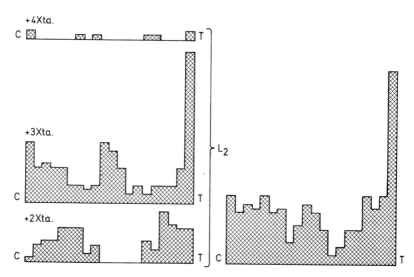

Fig. 2.6. Chiasma distribution for the L_2 bivalent of *Schistocerca gregaria*. In addition to the over-all distribution, the distributions for the bivalents with *2*, *3* and *4* chiasmata have been given separately. *C* = centromere; *T* = telomere. (After HENDERSON, 1963)

for any variance ratio. ROWLANDS (1958), observed somewhat lower frequencies of chiasmata in his *Vicia faba* material than MAEDA (1930) but again found deviations from a Poisson distribution with variances that were considerably smaller than the mean, indicating a coincidence smaller than 1, and significant interference (Table 2.4B). It must be noted that coincidence estimates derived by this method are not the same as those made on the basis of more direct cytological or genetic approaches.

With such comparisons, the information contained in the distribution of chiasmata in and over bivalents is not exhausted. It is possible to test where the deviation from a Poisson series starts. It may be assumed that a single chiasma in a bivalent does not experience interference, i.e. the first chiasma is formed without restriction. It is also conceivable that two chiasmata in large chromosomes, although perhaps maintaining a certain distance, do not suffer a reduction in frequency as a result of interference. If the first chiasma is indeed free to be formed, it belongs to the same frequency distribution as the class in which no chiasma is formed at all, where interference cannot play a role. In a Poisson distribution, the expectation for class 0 (no chiasmata) is $e^{-\mu}$, that for class 1 (one chiasma) is $\mu \cdot e^{-\mu}$. If these two classes belong to the same distribution, they must have the same average chiasma frequency per bivalent, μ.

The data of MAEDA (1930) and ROWLANDS (1958) are not satisfactory for this approach: the numbers in the low-chiasma classes are too small. In FOGWILL'S (1958) *Lilium longiflorum*, however, enough bivalents with 0,1, 2, 3, and 4 chiasmata were observed (three clones pooled; male meiosis). The 10S bivalents, all of similar size, were pooled (cf. Table 2.12), and the following distribution was obtained:

Chiasmata per bivalent:	0	1	2	3	4
Number of bivalents:	7	158	302	115	17
Frequency:	0.0117	0.2638	0.5042	0.1920	0.0284

The mean is 1.9616 (599 chromosome pairs); for the three clones the ratios between mean and variance were 0.288, 0.332 and 0.425 respectively, giving interference values of 0.712, 0.668 and 0.575.

The frequency of class 0 is $e^{-\mu}=0.0117$; the corresponding mean (μ) is 4.448. This is more than twice the observed mean. If a single chiasma is also free of interference, bivalents with one chiasma may be expected to have a frequency of $4.448\ e^{-4.448}=0.0516$, which they certainly do not. This class apparently is over-estimated, and it must contain a number of bivalents which might potentially have had two chiasmata, but which as a result of interference have realized only one. Therefore, even though at times up to four chiasmata can be formed simultaneously in one bivalent, as a rule even two is too much to be realized consistently. Since the number observed (7) in class 0 is low, the same comparison can be made more reliably on the basis of the frequency of class 1. From $\mu\cdot e^{-\mu}=0.2638$, by iteration $\mu\approx2$ results. When from this μ the expected frequency of class 0 is computed, a value of 0.1344 is obtained, or 80 chromosome pairs in a total of 599. This is significantly more than 7. This result can partly be explained by chiasma localization as well as by interference, but certainly not entirely.

There is a second approach to detecting and estimating interference. It amounts to comparing observed frequencies of simultaneous chiasma formation in *specific segments* of a chromosome with frequencies expected on the basis of independent formation of chiasmata. When the segment considered can contain more than one chiasma, interference is best described as a correlation between the segments in respect to chiasma formation. A positive correlation indicates negative interference, or, interference is negative when high (or low) numbers of chiasmata occur simultaneously more frequently than expected on the basis of randomness. A negative correlation results from positive interference.

For unmarked, normal, bivalents the only segments to be distinguished are the two arms of (sub-) metacentric chromosomes. There are a number of quantitative reports on chiasma formation in the two arms of a bivalent, HARTE'S (1956) data on *Paeonia tenuifolia* being a good example. She does not give the actually observed averages for the two arms, but does present correlations. The analysis is most straightforward for the sub-metacentric (Sm) and for the sub-acrocentric (St) chromosomes, where the two arms can be distinguished at metaphase (Table 2.5). The correlation is derived from a simple regression analysis.

When the two arms cannot be distinguished, a normal regression analysis is not possible, but there are two alternative approaches. One is based on estimating intra-class correlations and is not restricted to *pairs* of observations as in the case of the two arms of a bivalent, but can be applied for correlations between several observations in one class. The general formula for the estimate of this correlation coefficient (SNEDECOR and COCHRAN, 1967) is: $r_1 = (s_b^2 - s_w^2) \cdot s_b^2 + (n-1) s_w^2$ in which s_b^2 is the variance between classes and s_w^2 the variance within classes, while n is the number within a class. The other approach is estimating a correlation coefficient along lines similar to that followed for the common regression analysis. Since the two arms cannot be distinguished (and this is quite a common situation) each pair of values is entered twice, once with one arm considered to be left and the other right, the second time reversed. It leads to a symmetrical double table in which the sums and the variances of the two columns are equal. Now the equation for the estimate of the correlation coefficient is

$$r = \frac{2\Sigma(X - \bar{X})(X' - \bar{X})}{\Sigma(X - \bar{X})^2 + \Sigma(X' - \bar{X})^2},$$

where the sums include all pairs and $\bar{X}$ is the mean of all observations (SNEDECOR and COCHRAN, 1967). The two estimates are slightly different. The latter approach

Table 2.5. Regression coefficients for the relation between the chiasma frequencies in the two arms of the chromosomes of *Paeonia tenuifolia*. The M and m chromosomes have approximately equal arms and for the calculation of the regression coefficients symmetric double tables were constructed. For the Sm and St chromosomes the regression could be calculated both ways. Long arm $= x$; short arm $= y$ (HARTE, 1956)

Preparation	M-chrom.	Sm-chrom.		St-chrom.		m-chrom.
	$b_{yx} = b_{xy}$	b_{yx}	b_{xy}	b_{yx}	b_{xy}	$b_{yx} = b_{xy}$
1	−0.034	+0.147	+0.239	−0.079	−0.212	−0.203
2	−0.164	−0.145	−0.123	−0.139	−0.409	+0.015
3	−0.143	+0.318	+0.608	−0.064	−0.289	−0.124
4	−0.083	+0.188	+0.277	−0.228	−0.578	+0.051
5	−0.089	−0.106	−0.130	−0.087	−0.381	−0.016
6	+0.526	−0.722	−0.286	−0.259	−0.563	−0.204
7	+0.098	−0.049	−0.059	−0.162	−0.177	−0.231
8	−0.523	−0.025	−0.043	−0.065	−0.240	−0.238
9	+0.033	+0.038	+0.052	−0.125	−0.522	−0.119
10	+0.143	+0.277	+0.434	−0.009	−0.033	−0.282
Total	−0.043	+0.069	+0.094	−0.141	−0.469	−0.096
Significance of total regr.	−	(+)	(+)	+++	+++	+++
					r = −0.258	
Significance of difference between prep.	+	++	++	−	−	+++
Number of bivalents	1095	547		547		2091

was used by HARTE (1956) for estimating the regression for the two arms of the large and small metacentric bivalents in Paeonia (Table 2.5).

These regressions have been given in some detail in order to show a phenomenon of apparently general importance: a great variation between different individuals and even between different groups of cells within an individual, in which the different chromosomes appear to have a definite individuality. Whereas the St chromosome consistently shows negative regressions between the two arms (a statistically significant ($P<0.001$) positive interference across the centromere), in the other chromosomes both negative and positive regressions can be observed, which may occasionally be significant, but which on an average are not. Only in the case of the two m-chromosomes, the average regression for all preparations was significantly negative. The *differences* between the regression coefficients of the separate preparations were significant for all chromosomes, except St, which was homogeneous.

Most other reports are not as detailed as HARTE'S (1956), but some contain special methods of analysis and interesting quantitative information. CALLAN and MONTALENTI (1947) studied interference between the two arms of the chromosomes of two species of mosquitoes and found clearly different patterns. In *Culex pipiens* the three bivalents are almost exactly metacentric, two are of the same size and the third is slightly smaller. Each arm had 0 or 1 chiasma, and rarely two. The latter were not considered, and the situation is much the same as in the case of chromosomes where merely the association frequency of arms is noted, and not the frequency of chiasmata (Section 2.1.2). Since diplotene and metaphase chiasma frequencies did not differ significantly, the latter are as representative as the first. In their analysis of between-arm interference the authors started from the assumption that all arms had an equal chance of forming a chiasma, probability p. No chiasmata: $1-p=q$. For two chiasmata in a bivalent (one in either arm) the expectation is p^2, for a chiasma in one arm, none in the other: $2pq$, and for no chiasmata (the chromosomes are univalent): q^2. In one specimen of Culex 110 bivalents were studied, which had a total of 115 chiasmata. Then $p=115/220=0.523$ and $q=0.477$. The expected number of bivalents in which both arms had a chiasma was $0.523^2 \times 110 = 30.05$; one arm with and the other without chiasmata: $2 \times 0.5227 \times 0.4773 \times 110 = 54.88$; two univalents: $0.4773^2 \times 110 = 25.06$. The observed values were: 10, 95 and 5. The difference is highly significant, and since the classes with two and no arms having chiasmata are deficient, the conclusion must be that there is strong across-centromere interference. For all 10 specimens studied the same trend was observed, but there was some variation in the degree of interference. When a is the number of bivalents with one chiasma in each arm; $2b$ the number with only one chiasma, and c the number of univalent pairs (no chiasmata), the frequency of a chiasma in one arm, in absence of a chiasma in the other equals $p_1 = \dfrac{b}{b+c}$. The frequency of a chiasma in one arm in presence of one in the other is $p_2 = \dfrac{a}{a+b}$. The ratio $\dfrac{p_2}{p_1}$ is $\dfrac{a(b+c)}{b(a+b)}$, and the *index of interference* $1 - \dfrac{a(b+c)}{b(a+b)}$. This index is 0, when there is no interference, i.e. when $p_1=p_2$.

With complete interference, the index is 1, because $p_2=0$. In the example given, the highest index was 0.827, in the series of 10; the lowest was 0.414. There was, as expected, an inverse correlation with chiasma frequency.

CALLAN and MONTALENTI (1947) also considered the possibility that interference was absent, but that the difference between observed and expected (on basis of equal arms) frequencies of configurations were due to differences between arms. They followed a procedure very closely resembling that considered in Section 2.1.2, and came to the conclusion that in this case for specimen 1 a "long" arm with 0.9497 times a chiasma and a short arm with 0.0957 times a chiasma would have to be assumed. It was concluded that such a difference would be unrealistic, and that the assumption of across-centromere interference was valid. This is quite acceptable, but the fact that both solutions are in fact possible, shows the difficulty arising when limited data are used to estimate several parameters. This dilemma will be reconsidered further on in this chapter.

The Culex material made it possible to collect more specific additional information on interference between the arms. For this purpose, the location of chiasmata in the arms was scored, P being a proximal chiasma (located in the $\frac{1}{2}$ arm nearest the centromere), D being a distal chiasma. By 0, absence of chiasmata was indicated. The number of bivalents without chiasmata (univalent pairs) was 00, that of bivalents with one arm without chiasmata and the other with a distal chiasma by $0D$, etc. Then, given no chiasma in one arm (total frequency $00+\frac{1}{2}0P+\frac{1}{2}0D$), the chance of no chiasma in the other is

$$\frac{00}{00+\frac{1}{2}0P+\frac{1}{2}0D}.$$

The chance of a proximal chiasma in the other arm is

$$\frac{\frac{1}{2}0P}{00+\frac{1}{2}0P+\frac{1}{2}0D},$$

and a distal chiasma in the other arm:

$$\frac{\frac{1}{2}0D}{00+\frac{1}{2}0P+\frac{1}{2}0D}.$$

Similarly, given a proximal chiasma in one arm, the chance of no chiasma in the other is

$$\frac{\frac{1}{2}0P}{PP+\frac{1}{2}0P+\frac{1}{2}PD},$$

a proximal chiasma in the other:

$$\frac{\frac{1}{2}PP}{PP+\frac{1}{2}0P+\frac{1}{2}PD},$$

and a distal chiasma in the other:

$$\frac{\frac{1}{2}PD}{PP+\frac{1}{2}0P+\frac{1}{2}PD}.$$

The last three formulae in this set of nine refer to a distal chiasma in one arm and no chiasma in the other:

$$\frac{\frac{1}{2}0D}{DD+\frac{1}{2}0D+\frac{1}{2}PD},$$

a proximal chiasma in the other:

$$\frac{\frac{1}{2}PD}{DD+\frac{1}{2}0D+\frac{1}{2}PD},$$

a distal chiasma in the other:

$$\frac{\frac{1}{2}DD}{DD+\frac{1}{2}0D+\frac{1}{2}PD}.$$

The results for a second specimen of *Culex pipiens* are summarized as follows:

Chance of chiasma in other arm

		O	D	P
Chiasma in	O	0.022	0.769	0.209
one arm	D	0.422	0.482	0.096
	P	0.514	0.432	0.054

In general the chiasmata are distally located, but it is clear that the chance for a proximal chiasma to occur decreases from 0.209 to 0.096 to 0.054 when in the other arm no chiasma, a distal chiasma and a proximal chiasma occurs. Three more specimens were analyzed following the procedure, and all showed the same trend.

Theobaldia longiareolata Macq. is closely related to Culex. It also has three chromosome pairs, and again one is shorter than the other two. One of the long pairs is slightly submetacentric but its arms cannot be distinguished at meiosis. In CALLAN and MONTALENTI'S (1947) analysis, the short bivalent had an average chiasma frequency of 1.953 ± 0.041 and seldom had more than one chiasma in either arm. The two larger bivalents, however, could have up to three chiasmata in an arm and were strikingly different from the Culex bivalents. They had a mean chiasma frequency of 2.911 ± 0.069 in a total of 170 bivalents. The short bivalents formed rings almost without exception, showing no indication of interference. The number of cells analyzed was too small for a test of the distribution of chiasma position in the large bivalents, but PP

bivalents seemed to be in excess rather than showing a shortage. Expectations for different combinations of numbers of chiasmata in the two arms (based on independence of the arms and equal chances for each arm for each number of chiasmata) were calculated and compared with the observed numbers of these combinations (Table 2.6). There was perfect agreement and clearly there is no interference across the centromere. For the calculation of the expected frequencies first the frequencies of arms with 0, 1, 2 and 3 chiasmata are derived from the first and second columns of Table 2.6. These frequencies are 0.0206, 0.5265, 0.4235 and 0.0294 respectively. The expected number of bivalents with 0 chiasma in one and 1 chiasma in the other, for instance, is calculated as $2 \times 170 \times 0.0206 \times 0.5265 = 3.672$. Although not calculated by CALLAN and MONTALENTI (1947) the average number of chiasmata per arm (1.462) can be used to calculate the expected (Poisson) distribution: the probabilities are 0.2318 for 0, 0.3388 for one, 0.2476 for two, 0.1210 for three and 0.0589 for four chiasmata. The difference with the observed frequencies given above is considerable and must be due to interference. The ratio between average and variance gives a quantitative measure for this interference. For the entire bivalent it is $\dfrac{2.924}{0.8048} = 3.633$. Compared to Culex, with an average of 1.0454, a variance of 0.1296 (ratio 8.066) the interference in Theobaldia is relatively weak. This weak interference is reflected in the relatively high realized chiasma frequency, and the failure to detect interference across the centromere may merely be an aspect of low overall interference. In Culex interference is much stronger, is detectable across the centromere, and is one of the reasons of the low chiasma

Table 2.6. The distribution of chiasmata in the arms of the two large metacentric bivalents of *Theobaldia longiareolata*. Expectations based on independence and equal chances of chiasma formation in both arms. (From CALLAN and MONTALENTI, 1947)

Bivalent type		Observed no. of bivalents	Expected
Chiasma in one arm	other arm		
0	0	0	0.071
0	1	5	3.672
0	2	2	2.965
0	3	0	0.207
1	1	51	47.124
1	2	69	75.810
1	3	3	5.270
2	2	33	30.490
2	3	7	4.233
3	3	0	0.147
		170	169.989

Average per bivalent 2.924; per arm 1.462
Variance per bivalent 0.8048
Ratio average/variance 3.633

frequency. It is not the only reason, since the class of 0 chiasmata per chromosome pair is considerable, and not, or only slightly, affected by interference.

The combinations of high chiasma frequency with no detectable across-centromere interference, and low chiasma frequencies with across-centromere interference are not uncommon. In *Fritillaria chitralensis*, BENNETT (1938) found no significant (slightly negative, -0.1516) correlation between the two arms of a sub metacentric bivalent, and relatively high chiasma frequencies. In Dasyurus, KOLLER (1936) counted an average of 1.6 chiasmata per bivalent and a strong chiasma interference, whereas in Sarcophilus with 2.4 chiasmata per bivalent, interference was weak. The importance of the presence of a proximal chiasma for the expression of across-centromere interference was indirectly demonstrated by SYBENGA and DE VRIES (1972) in rye. This will be discussed in 2.1.2.

The correlation between low chiasma frequency with across-centromere interference is not strict: when the cause of the low frequency is not interference but some other factor (distal localization; pairing failure around the centromere as a result of distal pairing initiation), there is no basis for across-centromere interference as a large segment aroung the centromere simply is not available for chiasma formation. Actually, incidental pairing failure around the centromere may be a cause of *negative interference* across the centromere: when there is pairing, there will be a tendency for a chiasma to be formed on both sides simultaneously, even when (relatively weak) positive interference keeps the two well separated. When pairing fails, there will be no chiasma on either side. One chiasma on one side and none at the other will be relatively uncommon. Another cause of apparent negative interference is variability in chiasma formation between cells: a cell with a relatively high chiasma frequency will have a relatively great chance of having a chiasma on both sides of the centromere. A cell with a low chiasma frequency will relatively frequently have no chiasma on both sides. Both factors may play a role in *Trillium erectum* where NEWCOMBE (1941) found negative interference across the centromere. Chiasmata were concentrated around the centromere, and their frequency was relatively high, indications of low (positive) interference in this region. NEWCOMBE'S approach is a special variant of the analysis of interference. It is related to the methods commonly employed in genetic experiments and relates the occurrence of a chiasma in one specific segment to that in another. Chiasma formation in segments of 1 micron on both sides of the centromere was recorded (Table 2.7A) and the coincidence estimated. For instance, in 48 bivalents of chromosome B, 23 had a chiasma in the proximal micron of one arm, and 29 had one in the proximal micron of the other arm. The expected number of bivalents with chiasmata simultaneously in both segments was $\frac{23}{48} \times \frac{29}{48} \times 48 = 13.9$. Observed were 20, and the coincidence c is 1.43. Since interference is 1-coincidence, it is negative: -0.43, in spite of the close proximity of the two segments.

Interference within Arms. The same Trillium material used for the analysis of across-centromere interference (NEWCOMBE, 1941) was used for studying within-arm interference. The bivalents were divided in units of appr. 1 micron along their entire lenght, and the chiasma frequencies for each unit were recorded. Each type of bivalent was always divided into the same number of units, and since there was some variation in length between cells, the units were somewhat longer in some cells than in others. Contrary to what was found for the centromeric region, adjacent units tended to have low coincidence values, but these increased with distance, and after two microns approached 1. This was more pronounced, of course, with high than with low chiasma frequencies. Pooled data for the five bivalents are given in Table 2.7B. There is some correspondence with the analyses of HENDERSON (1963) and FOX (1973), see Sections 2.1.1.1, 2.1.1.2; Figs. 2.6, 2.8; Table 2.2.

Table 2.7(A). Distribution of chiasmata in the five bivalents of *Trillium erectum* (4 slides, 12 cells each). The divisions used were obtained by dividing the arms into a given number of regions in such a manner that in an arm of average length these would be as close as possible to one micron (NEWCOMBE, 1941)

Chromo-some	Number of biv.	Regions in left arm										Centromere			Regions in right arm								
		10	9	8	7	6	5	4	3	2	1	1	2	3	4	5	6	7	8	9	10	11	12
A	48											16	6	5	9	12	10	99	17	4	5	5	14
B	48							5	3	2	23	29	13	11	10	11	14	16	8	14	10		
C	46						7	7	2	8	29	28	12	7	5	14	7	10					
D	48				4	13	9	6	17	10	34	38	9	11	12	20	14	12	10				
E	48	12	14	4	16	15	7	7	5	0	38	37	4	7	7	15	17	13	14	7	8	18	

Table 2.7(B). Coincidence of chiasmata within chromosome arms of *Trillium erectum* for segments separated by different distances. (From data of NEWCOMBE, 1941)

Distance in microns	Chiasmata in			Number of regions considered	Coincidence	Interference
	First region	Second region	Both			
	a	b	x			
0	715	725	91	3100	0.544[b]	0.556
1	643	639	139	2672	0.904[a]	0.096
2	568	581	136	2244	0.925[a]	0.075
3	488	486	129	1816	0.949	0.051
4	403	354	110	1436	1.107	−0.107
5	315	277	76	1102	0.960	0.040
6	223	177	54	768	1.051	−0.051

[a] In some slides significantly deviating, in others not deviating from 1.
[b] Consistently significantly deviating from 1.

2.1.1.3 Interchromosome Effects

Already in 1933 MORGAN et al. discovered that in *Drosophila melanogaster* reduction of genetic crossing-over (later found to be due to inversion heterozygosity) in some chromosomes resulted in an increase in crossing-over in specific segments of other chromosomes (for a review up till 1950, see SCHULTZ and

REDFIELD, 1951). This is a unilateral effect. Reciprocal effects between different recognizable chromosomes were observed by DARLINGTON (1933) in rye, when he compared the chiasma frequency (per cell) in a B-chromosome bivalent with that in the remainder of the bivalents. There was a clear negative correlation (positive interference): when the B-bivalent had a high chiasma frequency, the other bivalents had a low frequency, and *vice versa*.

SAX (1935) estimated the correlation between chiasma frequencies in the single long bivalent and the five short bivalents in *Vicia faba*. Since the two types of bivalent could easily be distinguished, the analysis was relatively simple, and when no correlation was found, it was concluded that the long and the short bivalents were independent in respect to chiasma formation. The analysis was repeated by ROWLANDS (1958) who found a whole range of correlation coefficients when studying a series of different plants (Table 2.4 b). All were negative, some significantly so. HARTE (1956), in the material of Paeonia already described above, calculated the correlation coefficients between pairs of different chromosomes, and also between individual chromosomes and the rest of the genome. The between-chromosome correlations are given in Table 2.8. The great variation in correlations is striking. Some are significantly negative and can be considered due to some form of positive interference, others, however, are positive.

Table 2.8. Correlation coefficients for chiasma frequencies of bivalents within cells and regression coefficients for chiasma frequencies of a bivalent and the rest of the genome within cells in *Paeonia tenuifolia*. Cf. Table 2.5 (HARTE, 1956). Two out of 10 preparations

Prepara-tion	M-chromosome vs.				Sm-chromosome vs.			St-chrom. vs.	
	Sm	St	$m_1 + m_2$	regression rest genome	St	$m_1 + m_2$	regression rest genome	$m_1 + m_2$	genome rest regression
I	−0.171	−0.066	−0.203[a]	−0.512[a]	+0.057	−0.207[a]	−0.422[a]	−0.222[a]	−0.448[a]
IV	−0.002	+0.052	+0.063	+0.156	+0.012	+0.033	+0.044	+0.031	+0.112
Total	−0.097	+0.023	−0.029	−0.128	−0.004	−0.074	−0.215	−0.125	−0.241

[a] Significant.

In *Endymion non-scriptus*, where five of the eight bivalents can be recognized at meiosis, ELLIOTT (1958) analyzed the relations between bivalents and found both positive and negative correlations, but only few were significant. Although some trends could be detected, irregularity predominated. Correlations tended to be significant with increasing variances.

A special statistical variant of the analysis of between-chromosome effects (intra-class correlations), was first applied by MATHER and LAMM (1935) in their study on interactions between rye chromosomes. The different bivalents in a cell could not be distinguished, and intra-cell variance (a measure of the variation between different bivalents in respect to chiasma frequency) was compared with inter-cell variance. Details of the calculations are given by MATHER (1965): the results and analysis of variance are shown in Table 2.9. When the intra-cell variance is the greater of the two, there is more variation between the different bivalents of a cell than between identical bivalents in different cells. This variation between different bivalents within a cell has two components: intrinsic differences between bivalents in respect to chiasma formation (for instance, longer bivalents may have more chiasmata) and a component due to interdependence of bivalents. Since the chromosomes of rye are practically identical in size, MATHER and LAMM (1935) considered the first component to be negligible. Interdependence between bivalents may be expressed as a positive correlation: all have either relatively high or relatively low numbers of chiasmata simultaneously, and then the within-cell variance is small compared to the between-cells variance. Or the correlation may be negative, as when some bivalents have a high, others have a low number of chiasmata. Then the within-cell variance is the greater. It is clear that with excessive variation between cells, due, for instance, to inhomogeneity in the tissue, the within-cell variance, although in itself large as a result of inter-chromosome effects, will appear to be relatively small. Then the interference between bivalents is not detected, or even converted into an apparent positive correlation. That, on the other hand, negative correlations may be induced spuriously by inhomogeneities *within* cells was shown by SYBENGA (1967a) in translocation heterozygotes of rye, which will be discussed in Section 2.2.4.4. Other forms of intra-cell inhomogeneity are described by REES (1961).

The analysis of interchromosome effects in outbred population plants (MATHER and LAMM, 1935) was extended by LAMM (1936) to inbred lines of rye. It appeared that the results were very variable: besides a few, not clearly significant negative correlations, several positive correlations appeared. LAMM attributed this to a breakdown of the normal control mechanism of crossing-over. It accompanied a general reduction of the overall chiasma-frequency.

Table 2.9. Frequency of chiasmata in rye, and the analysis of variance (MATHER, 1965; MATHER and LAMM, 1935)

| | Chiasmata per bivalent | | | | |
	1	2	3	4	Total
No. of bivalents	4	150	89	2	245

| | Chiasmata per nucleus | | | | | |
	14	15	16	17	18	Total
No. of nuclei	1	3	12	14	5	35

Item	Sum of squares	N	Mean square	Variance ratio
Between nuclei	4.3837	34	0.1289	2.3747 P 0.001
Within nuclei	64.2857	210	0.3061	
Total	68.6694	244		

In *Vicia faba*, ROWLANDS (1958) applied this method (of comparing variances) to the ten small, acrocentric bivalents which are morphologically very similar. Examples of four genetically different plants showed considerable variation (Table 2.10). *Vicia faba* is partially allogamous, and a direct relation with inbreeding is not easily detected, nor is it in most other material where variation in inter-chromosome relations has been established.

Table 2.10. Comparison of intra- and inter-cell variance as an indication of correlation of chiasma frequency for the small (m) bivalents of *Vicia faba*. Last columns: interclass correlation coefficients for the large (M) bivalent and the sum of the small (m) bivalents (ROWLANDS, 1958)

| Plant | Small (m) bivalents | | | M vs m bivalents corr. coeff. | |
| | Variance | | Variance ratio | | |
	Intra-cell	Inter-cell		r	N
J 0	0.4600	0.2584	1.8[a]	−0.2457	18
J 2	0.4800	0.3680	1.3	+0.2687	18
J 4	0.4600	0.7870	1.71[a]	+0.3010	18
J 8	0.4000	0.4940	1.24	−0.2452	18

[a] Significant at 5% level.

In the Endymion material of ELLIOTT (1958), when three different temperatures were tested in addition to field conditions, positive correlations predominated, but did not seem to be related to temperature. In Hyacinthus (ELLIOTT, 1958), between-bivalent (within-cell) variances varied with temperature more than the between-cells variances, but not in a specific direction.

With the purpose of detecting some system in these apparent irregularities, BASAK and JAIN (1963) in their analysis of interchromosome effects in *Delphinium ajacis* (two large and six small bivalents), introduced a further refinement. Correlation coefficients between the two groups of bivalents were negative and significant in all 28 plants studied. Significantly different were also the inter-nuclear and intra-group variances. The latter were larger, indicating a negative correlation, when it may be assumed that within the groups no significant differences in chiasma frequency per chromosome occurred. The plants were now grouped into categories with increasing chiasma frequencies. In the first category those with a range of 9–12 chiasmata were combined, in the second group those with a range of 10 to 13, the third with a range of 10 to 14 and the fourth with a range of 12–16 chiasmata (Table 2.11). The last two categories were represented by only two plants each, the first two by 9 and 15 respectively. Within each group of plants, cells with exactly 9 chiasmata, exactly 10 chiasmata, etc. were considered separately. Within these classes, the inter-nuclear variances, of course, were 0. Estimates for the intra-nuclear variances were calculated for each class of cells with any particular number of chiasmata. For instance, in the 9 plants of category 1 (range 9–12 chiasmata), the pooled intra-nuclear variance for those cells with all exactly 9 chiasmata was 0.0852, for those with

10 chiasmata 0.1979 etc. Two facts emerged: (1) The variances increased with increasing number of chiasmata, up to a certain point, after which they decreased. (2) The behavior was different for different categories, expecially in that the high-chiasma frequency plants showed an entirely different level of variances (Table 2.11). In itself a decrease in variance in cells with low and with high frequencies is expected, since the variance in these distributions has a definite relation with the mean, but it could be shown that this could explain only part of the difference. It is clear, then, that "the inter-chromosome negative correlation is a function of the number of chiasmata in a cell".

Another grouping of cells was carried out, all cells in each plant being classed (rather arbitrarily) into two groups, one with high, the other with low chiasma frequencies. In 23 out of 28 plants the correlations between the large and small bivalents were stronger (always negative) in the group with the high chiasma frequency than in the group with the low chiasma frequency. In 10 of these the difference was significant.

On the suggestion (originally by MATHER, 1936) that bivalents compete for chiasmata it was concluded by BASAK and JAIN (1963) that the competition would set in only above a certain minimum, and that it would be relaxed beyond a certain level. This would also explain why in the inbred rye studied by LAMM (1936) in which the average level of chiasma frequency was relatively low, negative correlations between bivalents were not common, whereas in normal population rye, they were the rule. As long as the mechanisms involved are not known, the term "competition" can be used only in a descriptive sense. The pattern observed points to an intricate system of regulation of crossing-over, not to be explained in simple terms.

A third type of analysis of interchromosome relations (in addition to the straight-forward correlation coefficients and the intra-class correlations) is based on the use of contingency tables. It will briefly be discussed in Section 2.2.4.2.2, in relation with translocations as chromosome markers. It is similar to the analysis of between-arm interference considered above for the B-chromosomes in rye (2.1.1.2).

Table 2.11. Intraplant, inter-cell group variation in intranuclear variance in *Delphinium ajacis*. Plants were grouped according to range of chiasmata per cell (group 1, plants 1–9 with 9, 10, 11 or 12 chiasmata per cell; group 2, plants 10–24 with 10, 11, 12, 13 chiasmata per cell; group 3, plants 25 and 26 with 10, 11, 12, 13 or 14 chiasmata per cell and group 4, plants 27 and 28 with 12, 13, 14, 15 and 16 chiasmata per cell; generally 100 cells per plant). Intra-nuclear variance calculated for groups of cells with identical numbers of chiasmata (inter-nuclear variance=0). BASAK and JAIN, 1963

| | | Pooled intra-nuclear variances in cell groups with: | | | | | | | | |
Group	Plants	9	10	11	12	13	14	15	16	chiasmata
1	1–9	0.0852	0.1979	0.2913	0.4745	—	—	—	—	
2	10–24	—	0.1749	0.2763	0.4236	0.3690	—	—	—	
3	25, 26	—	0.1850	0.3640	0.7587	0.5702	0.2417	—	—	
4	27, 28	—	—	—	0.1555	0.3305	0.5209	0.5745	0.6444	

2.1.1.4 Chiasma Localization

As shown above (cf. HENDERSON, 1963; FOX, 1973) interference can lead to
a restriction in respect to the randomness of chiasma distribution. The specific
patterns of distribution observed are similar to those found for genetic crossing-
over in Drosophila (MATHER, 1936).

Much of the work on chiasma localization (preference of chiasmata to be
formed in specific chromosomal segments) as a result of interference but also
as a result of other, here as yet unspecified causes, has been carried out on
first meiotic metaphase bivalents (cf. MATHER, 1940 b), when the observations
should be interpreted with some caution. According to the reliable report of
HEARNE and HUSKINS (1935) on Melanoplus, chiasmata start moving as soon
as the chromosomes of the bivalents separate. This movement decreases with
increasing contraction of the chromosomes, as a result of increasing adhesion
between the associated chromatids. The moment the movement stops depends
on the equilibrium between the forces causing movement, and chromatid stickiness. This equilibrium may be quite different for different organisms and may
even vary between individuals of the same species. In a number of organisms
chiasmata tend to be completely terminalized at metaphase, but this does not
imply their terminal formation. In special chromosomal constructions (trisomics,
translocations, for instance: Figs. 2.21; 2.23) it is clear that in a single terminal
attachment point several chiasmata must have collected, that have been formed
in well separated chromosome segments. For chiasmata observed in non-terminal
chromosome segments, it is clear that they have indeed been formed in regions
nearer the centromere, but at least some movement is probable.

There is considerable variation in observed chiasma localization (Fig. 2.7,
and MATHER, 1940). Even between the two sexes of the same species differences
can exist (Fig. 2.7, WATSON and CALLAN, 1963; VED BRAT, 1966; VOSA, 1972;
PERRY and JONES, 1974).

When the chiasmata are strictly terminally localized (distally in metacentrics,
distally or proximally in telocentrics), the organism, in spite of high chiasma
frequencies, is effectively free of recombination. This is only one genetic consequence
of chiasma localization. Another has been brought forward, for instance, by
BANTOCK (1972), who suggested that the observed linkage disequilibrium in
certain populations of Cepaea nemoralis can be correlated with a virtual absence
of chiasmata in certain chromosome regions containing genes with marked
morphological expressions (cf. PRICE, 1974 for other populations). Several authors
have studied, and speculated upon the relation between recombinational patterns
(chiasma frequency and distribution) and breeding systems (cf. ZARCHI et al.,
1972). The evolutionary significance of the regulation of recombination has
been reviewed by BODMER and PARSONS (1962).

In general, there seems to be a rather complex negative correlation between
the localization of chiasmata and that of heterochromatin (DYER, 1964; JOHN
and LEWIS, 1965) but this correlation is not the only basis of chiasma localization
other than that caused by interference. Genetic factors play a significant role
(see 2.1.1.5), but the mechanisms involved are largely unknown.

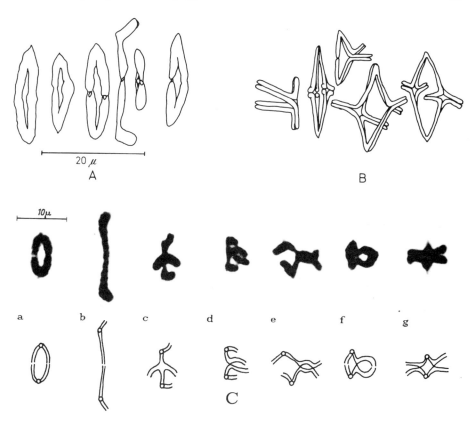

Fig. 2.7A–C. Chiasma distribution at M I in male (A) and female (B) meiosis in the newt *Triturus helveticus* ($n = 12$): distal in the male and unrestricted to proximal in the female. A selection of bivalents sketched after photomicrographs from WATSON and CALLAN, 1963 (cf. JOHN and LEWIS, 1965). (C) Distal chiasma localization in M I bivalents of normal rye (*Secale cereale*; a, b) and unrestricted to proximal in bivalents of derivatives of a hybrid between *S. cereale* and *S. turkestanicum* (c–g). (JONES, 1967)

2.1.1.5 Chiasma Frequency and Distribution in Relation to Internal and External Variables

The regulation systems of recombination at the level of crossing-over (or chiasmata) are influenced by numerous internal and external factors, the effects of which may vary from very slight modifications to drastic alterations and even complete breakdown. Sometimes the effects are straightforward and the analysis simple. In other cases the relations are more complicated and a sophisticated analysis is required. Since the basis of the regulation of recombination is genetic, first the genetic variation of chiasma frequency and distribution will be considered. From genetic experiments, for instance on the fungus *Schizophyllum commune* (SIMCHEN and STAMBERG, 1969) it is clear that the genetics of recombination is complex. Large numbers of genes are involved, some having an overall effect, others being site-specific. A similar complexity has not yet appeared from chiasma studies, but there are definite parallels. The approach has been diverse, varying from studies of monofactorial effects and polygenic variation to the analysis of interspecific hybrids and their derivatives. Monofactorial effects are usually rather drastic (they would not have been detected if they were not) and tend to be sensitive to environmental variation. When the recessive alleles effect a pronounced decrease in chiasma frequency the genes are referred to as "desynaptic". In PRAKKEN'S (1943) monofactorial desynaptic rye, not only frequency, but also distribution was modified. Numerous comparable, usually less detailed reports have been published. When chromosome pairing is affected primarily, the final effect may be similar, but the regulation system is not the same (cf. 3.1.4, Table 3.7).

One of the situations in which polygenic effects are encountered is in inbreeding programs. Inbreeding usually results in a decrease in chiasma frequency (LAMM, 1936; REES, 1955; SYBENGA, 1958), often accompanied by an increase in the variance within individuals, in spite of the fact that variance and mean of chiasma frequencies may be expected to be statistically correlated. This increased variance is considered to be an aspect of a general relaxation of control. An actual genetic analysis is usually not carried out. LAMM (1936) and SYBENGA (1958) described inbreeding depression in rye after several generations of selfing, starting with population plants. REES (1955) crossed two inbred lines of rye and recorded chiasma frequencies and their pattern of variation in the parents, the F_1, F_2 etc. The polygenic nature of the inheritance was concluded from the fact that all resulting lines had specific characteristics and that the parental types were not recovered. Breakdown of the genetic control of recombination upon inbreeding in naturally outbreeding material was demonstrated not only by an increase in within-individual variances, but also by alterations in localization patterns. REES (1955) described the regular occurrence of bivalents with proximally located chiasmata, which were extremely infrequent in normal material. When, as was suggested, these were (often) specific bivalents, this points to site-specific chiasma control.

Interspecific hybrids and their derivatives usually have reduced chiasma frequencies. The reason may be insufficient homology, but also, and independently, genetic imbalance resulting in the functional imperfection of the biochemical machinery. Genetic imbalance may also result in an upset of the regulating system without drastic reduction of the frequency of chiasmata as such, especially when the species involved are relatively closely related. An example is given by JONES (1967, cf. Fig. 2.7 C), of hybrid derivatives of the related species *Secale cereale* and *Secale turkestanicum*. One deviant type was studied in detail. Whereas in the normal genotypes chiasmata within bivalents were localized distally (Table 2.12 A) and between bivalents non-randomly distributed (Table 2.12 B), the abnormal genotype had less restricted chiasma distributions within bivalents (Table 2.12 A) and perfect random distributions of chiasmata between bivalents within cells (Table 2.12 C). Similarly, the distribution between cells was highly non-random in the normal, and completely random in the abnormal genotype. It may be concluded that interference was strong in the normal genotype (mean-variance ratio 6.57) and practically absent in the abnormal type (mean-variance ratio 1.22). The localization was also reduced, but not as strictly as interference. This can be interpreted as meaning that localization has two components: one due to interference and distal chiasma initiation, the other intrinsic. As appears from Table 12A, the total chiasma frequency was reduced only slightly. This comparison, however, is in fact not valid, since in absence of interference the normal type might have had a much higher chiasma frequency than observed.

The genetics of the abnormal type was studied by analyzing the chiasmata-per-bivalent variance in normal and abnormal types, and in their F_1 s and in their F_2 s. The normal type was clearly dominant, but the F_2 did not segregate into distinct classes. Either a small number of genes with similar effects, or a major gene with modifiers must have been involved.

A later analysis of the genetics of chiasma regulation in rye (JONES, 1974) was based on analyses of variance. Within cells, bivalents could not be distinguished, but from previous studies it was clear that there is little difference between the bivalents. Then, the only components of variation available for analysis are the between-cells mean square ($\sigma^2 + n\sigma_c^2$; $p-1$ degrees of freedom) and the within cells mean square ($\sigma^2 + \sigma_i^2 + k_b^2$; $p(n-1)$ degrees of freedom). Here p is the number of cells in the sample; n the number of bivalents in the cell. For the components of variation, σ_c^2 represents the differences between cells, k_b^2 those between bivalents, σ_i^2 the interaction between cells and bivalents,

Table 2.12(A–C). Chiasma distribution in normal and abnormal genotypes of rye, 50 cells (with 350 bivalents) for each type. (JONES, 1967)

(A) Distribution within bivalents

Position	Normal	Abnormal	χ^2
Distal	546	199	63.75
Interstitial	30	132	109.10
Proximal	0	71	101.73
Total	576	402	274.58

(B) Observed and Poisson distribution of chiasmata per bivalent in normal genotype

Chiasmata per bivalent	0	1	2	3	4	4+
Observed	5	92	247	6	0	0
Poisson	62.30	107.50	92.79	53.38	23.05	10.98
χ^2	52.70	2.27	256.29	42.06	23.05	10.98

(C) Observed and Poisson distribution of chiasmata per bivalent in abnormal genotype

Chiasmata per bivalent	0	1	2	3	4	4+
Observed	92	125	93	33	7	0
Poisson	100.20	125.30	78.37	32.68	10.46	2.99
χ^2	0.671	0.007	2.731	0.032	3.093	

and σ^2 the error. In a detailed analysis of five rye genotypes in which one specific bivalent was marked and could be recognized, and in addition one random bivalent per cell in an independent sample of cells was scored, the relative contributions of the variance components could be estimated using weighted least squares methods (Table 2.13). Apparently, the contribution of σ_c^2 and $(\sigma_i^2 + k_b^2)$ is statistically of little significance, indicating that between-genotype variation must be mainly a matter of random (error) variation. This means that the degree of control over chiasma formation is the main component of genetic variation.

Table 2.13. Components of chiasma frequency variation estimated from five different rye genotypes (JONES, 1974)

Genotype	σ^2	$\sigma_i^2 + k_b^2$	σ_c^2
J 21	0.4609 ± 0.1010 (71%)	0.1863 ± 0.1100 (29%)	
J 22	0.2796 ± 0.0693 (68%)	0.1198 ± 0.0742 (29%)	0.0137 ± 0.0173 (3%)
J 23	0.4531 ± 0.1122 (96%)	0.0068 ± 0.1153 (1%)	0.0136 ± 0.0245 (3%)
J 24	0.4362 ± 0.1037 (95%)	0.0162 ± 0.1077 (4%)	0.0048 ± 0.0196 (1%)
J 25	0.5006 ± 0.0376 (88%)		0.0649 ± 0.0268 (12%)

In an analysis of thirty different genotypes (ten plants of a commercial winter rye variety, line R; 10 plants of a semi-inbred hybrid derivative involving *Secale vavilovii* and *S.cereale*, line H; and 10 plants of an F_2 involving the distributional mutant mentioned above, line J) the correlations between the between-cell mean squares and the within-cells mean square per plant were highly significant. Per plant 50 PMC's were used. For the three lines together $r = 0.946$, $P < 0.001$; for line R, $r = 0.862$; $P < 0.01$; for line H, $r = 0.844$; $P < 0.01$ and for line J, $r = 0.748$; $P < 0.02$. Between them, the lines, and also the plants within the lines, were highly variable. Since the two mean squares have only σ^2 in common, the correlation must be due to variation in this error component. This again shows the importance of the degree of control (error variance) in genetically determined chiasma regulation.

In an additional experiment, the relation between frequency and position of chiasmata was analyzed. Twenty plants of an F_2 involving the distributional mutant (the J-line, see JONES, 1967) were used. For each plant the total between-bivalents mean square was calculated from 50 cells, and the position of the chiasmata in the bivalent (proximal, interstitial, distal) was noted. The correlation between mean-square and the proportion (transformed to angles) of non-distal chiasmata was highly significant: $r = 0.900$; $P < 0.001$. Since the probability of interstitial and proximal chiasmata was not correlated to chiasma frequency in the bivalents in which they occurred, the correlation between (error) mean square and lack of localization cannot be explained by the necessity of filling more proximal locations when the distal locations are occupied in high-chiasma bivalents: monochiasmate bivalents have their proportional share of non-distal chiasmata. In these genotypes, therefore, relaxation of control leads to the formation of non-distal chiasmata, as was also observed by REES (1955) in inbred material. JONES (1974) concluded that there must be a relatively simple genetic system of chiasma control. It may have two steps, as also suggested for Drosophila meiotic mutants; the number of potential exchange sites is controlled, and also, independently, the actual probability of exchange at any site. Both systems apparently have several controlling genes.

Effects of the *environment* have already been mentioned (e.g. ELLIOTT, 1958), and on this subject too, many more reports have been published. For an older review of temperature effects see WILSON (1959); for a review of the effects of ionizing radiations see WESTERMAN (1967).

The effect of *additional* or *lacking* chromosomes is variable. MATHER (1939) ascribes a generally (although not universally) observed increase in triploids to an expansion of the biochemical machinery, while the number of effective pairing sites in which the chiasmata are to be deposited is not increased. In the tetraploid the level is usually back to that in the diploid (the number of pairing sites is doubled also) or even lower. Addition or loss of individual chromosomes usually leads to gene unbalance which may result in a reduction of chiasma frequencies. Addition of B-chromosomes has a variable effect: in cultivated rye JONES and REES (1967) found no significant effect on the mean chiasma frequency in A chromosomes, but an increase in variance. In wild populations of rye ZEČEVIĆ and PAUNOVIĆ (1969) observed an increase in the mean chiasma frequency in the presence of B-chromosomes. Comparable reports are available for other species.

In general, as shown in detail for genetic control, a change in frequency is correlated with a change in distribution (DARLINGTON, 1937, 1965) but this correlation is not a simple one. In the case of the B-chromosome of JONES and REES (1967), the total frequency was affected less than the distribution. According to HENDERSON (1969) alterations in crossing-over between specific loci (for instance detected in genetic experiments) are often merely due to changes in localization. A decrease in the segment analyzed is then accompanied by an increase elsewhere, which is not necessarily detected.

Correlations of chiasma frequencies with *chromosome length* indirectly involve interference and localization. Chromosome length can be determined at diplotene or at somatic metaphase. HEARNE and HUSKINS (1935) noted that in Melanoplus the long bivalents, which at diplotene were seven times the length of the shortest bivalent, had only 1.7 times the number of chiasmata. The medium long bivalents (approximately three times the length of the shortest) had only 1.03 times as many chiasmata. In another grasshopper, *Schistocerca gregaria*, HENDERSON (1963) and FOX (1973) found a somewhat different relationship (Fig. 2.8B). The shortest bivalents had exactly one chiasma each, independent of length, but for the longer chromosomes with over two chiasmata on an average, chiasma frequency increased linearly with length. On the basis of similar observations in *Locusta migratoria*, MATHER (1937) constructed a generalized graphic presentation of the relation between chiasma frequency and chromosome length (Fig. 2.8A). DARLINGTON (1937, 1965) noted very similar relations for several species. Irregularities, however, are common. HENDERSON (1963) explained the somewhat irregular behavior of the medium-size bivalents of Schistocerca by frequent simultaneous formation of a chiasma at both ends of the bivalents, which gave the longer medium-sized bivalents preferentially two chiasmata (Fig. 2.8B).

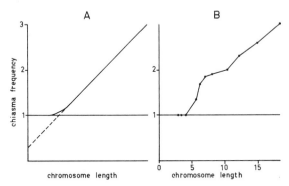

Fig. 2.8A and B. The bivalent length *vs.* chiasma frequency relationship. (A) Generalized according to MATHER (1937), mainly on the basis of observations on *Locusta migratoria*. (B) More exactly shown for *Schistocerca gregaria* (drawn after HENDERSON, 1963). There is a minimum of one chiasma, independent of length. The hump in (B) reflects a sharp increase in the frequency of medium size bivalents with two chiasmata, ascribed to frequent chiasma initiation near or at both ends

ELLIOTT (1958) compared metaphase chiasma frequencies in individual chromosome arms of *Endymion non-scriptus* (Table 2.3) with their somatic length. Although there was a general correlation, some exceptions occurred, in which longer arms had fewer chiasmata than shorter arms. In HARTE's (1956) *Paeonia tenuifolia* chromosomes, there was again a tendency for the long chromosomes to have more chiasmata, but the smallest (metacentric) chromosomes had more chiasmata than the larger sub-acrocentric and sub-metacentric chromosomes.

It has been shown (for instance by HENDERSON, 1963, but also by others) that specific bivalents tend to vary in relative length (for instance expressed as percent of the total genome) and that they tend to have more chiasmata, the longer they are. When BENNET and REES (1970) provided rye plants with different doses of phosphate, it appeared that with high doses chromosome size increased, as did chiasma frequency. Between cell variance (of chiasma frequencies) decreased. This suggests that the greater length is not a consequence of a higher number of chiasmata but rather that the reverse is true. This is also proposed by FOGWILL (1958) as a partial explanation of the differences between sexes in respect to chiasma frequency: in female cells the bivalents are larger than in male cells, and also have more chiasmata.

It is clear that complex systems of regulation exist, which are partly overall genotypic and partly chromosome-bound. When comparing the length-frequency relations of different species, it seems that interference and localization determine the differences rather than a basic level of chiasma frequency.

The relation between recombination and sex can best be studied in genetic experiments, as usually it is a tedious task to collect sufficient numbers of oocytes (in animals) or megasporocytes (in plants). A review of oocyte chiasma frequencies in mammals has been given by JAGIELLO (1975). Comparison with spermatocyte chiasma frequencies, when available, demonstrates the general trend of higher female chiasma frequencies. There are a few examples of direct cytological comparisons, one of which is by PASTOR and CALLAN (1952) on the turbellarian worm *Dendrocoelum lacteum*, another (for plants) by FOGWILL (1958) on Lilium (several species) and *Fritillaria meleagris*. The animal and plant data are comparable, as Dendrocoelum as well as the plants are hermaphrodites, and in both instances the frequencies of chiasmata in male and female cells have been taken from the same individual. This was not without complications in Dendrocoelum where meiosis in spermatocytes preceeds that in oocytes, and identical stages were not available in sufficient numbers in the same individual. Diplotene in oocytes had to be compared with metaphase in spermatocytes. PASTOR and CALLAN (1952) state that in the few diplotene stages studied, number and location of chiasmata corresponded with those at metaphase and therefore in this respect a difference in stage is not disturbing. One difficulty remained: in diplotene oocytes, twists in the bivalents could occasionally have been mistaken for chiasmata. In order to exclude errors due to such misclassification, in addition to chiasma numbers the numbers of bivalents with only one chiasma, which could be distinguished unequivocally, were determined (Table 2.14A). These comparisons are of considerable importance, since they demonstrate that the assumption of identity of male and female recombination frequency is not valid.

The mean chiasma frequencies in the spermatocytes of six individuals of Dendrocoelum varied between 9.3 ± 0.6 and 13.0 ± 0.3 (10 cells each). In the oocytes the range of the means was from 17.0 ± 0.6 to 22.0 ± 0.6. It is clear that the oocytes have a considerably higher chiasma frequency than the spermatocytes. The correlation ($+0.94$) between the spermatocyte and oocyte chiasma frequencies in the six individuals was highly significant: genetic as well as environmentally conditioned variation in the control of chiasma formation is expressed identically in both sexes but at a different base level.

The results for the plant species were very similar. Identical stages could be compared. The female frequencies were invariably higher and less variable than the male frequencies. Some of FOGWILL'S (1958) data are given in Table 2.14B. Separate analyses were given for the S and M bivalents. Temperature effects were studied, but appeared to be erratic. Interference does not seem to be significantly different. In some instances the female bivalents were decidedly larger than the corresponding male bivalents. This agrees with the relation between bivalent length and chiasma frequency (for the *same* bivalent in different cells) as observed in Schistocerca (HENDERSON, 1963; FOX, 1973). The cells, of course, were also larger. Two of the factors suggested by FOGWILL (1958) to explain the higher female chiasma frequencies are a difference in time available for pairing (more in females) and nutritional differences (better in oocytes and embryosac mother cells). For Allium (VED BRAT, 1966) and Tulbaghia (VOSA,

Table 2.14(A). Chiasma frequencies in spermatocytes and oocytes in the hermaphroditic worm *Dendrocoelum lacteum* (PASTOR and CALLAN, 1952)

Specimen	Cells	Spermatocytes				Cells	Oocytes			
		Chiasm. per cell	SE	% biv. with 1 chiasma	% biv. with >1 chiasmata		Chiasm./ cell	SE	% biv. with 1 chiasma	% biv. with >1 chiasmata
13	10	11.9	0.4	34.3	65.7	10	20.0	0.7	1.4	98.6
14	10	9.3	0.6	63.0[a]	34.3	10	17.0	0.6	5.7	94.3
15	10	12.9	0.5	22.9	77.1	10	22.0	0.7	0.0	100.0
19	8	12.1	0.4	26.8	73.2	10	20.0	0.8	0.0	100.0
21	10	11.7	0.4	37.1	62.9	9	21.4	0.8	0.0	100.0

[a] Two pairs of univalents.

1972) comparable results were obtained, but in rye (DAVIES and JONES, 1974) no difference between male and female chiasma characteristics were observed.

As shown above, not only the frequency, but also the localization of chiasmata can be different for males and females and terminal localization may be an even stronger restriction on recombination than reduced chiasma frequencies.

Sex differences, as far as they are expressed within an individual, as in hermaphrodites, represent a special form of intra-individual variation. Less special forms of this variation are also of very general occurrence. Whenever different groups of meiocytes are observed, they tend to have different mean chiasma frequencies and localization patterns. A well known example is the harlequin lobe of the testes of the axolotl, which is asynaptic. Such extremes are infrequent. In rye (REES and NAYLOR, 1960) and in barley (KÜNZEL, 1963), and in less detail in some other plants it has been observed that different parts of the same inflorescence or the same flower or even the same anther can systematically show differences in chiasma frequency. It is a common experience among cytologists to observe rare configurations in clusters instead of randomly distributed. Apparently all these cases reflect variations in the pattern of regulation

Table 2.14(B). Chiasma frequencies in pollen mother cells and embryosac mother cells in two species of Lilium (FOGWILL, 1958); 10 subacrocentric short (S) bivalents and two metacentric, larger (M) bivalents

Clone	Sex	Cells	Mean		Variance (bivalents)		Variance/ mean ratio	
			M	S	M	S	M	S
Lilium longiflorum								
3	♂	20	3.45	2.5	0.40	0.67	0.116	0.268
	♀	6	4.08	2.9	0.11	0.69	0.027	0.238
8	♂	20	2.77	1.79	0.71	0.76	0.256	0.425
	♀	7	3.28	2.4	0.95	0.73	0.290	0.304
9	♂	10	3.15	2.2	0.63	0.73	0.200	0.332
	♀	4	3.37	2.5	0.76	0.65	0.226	0.260
Lilium martagon								
1	♂	13	3.77	2.89	0.78	0.93	0.207	0.322
	♀	6	4.33	3.3	0.49	0.61	0.113	0.185

of chromosome behavior, partly conditioned by variations in the biochemical properties of different parts of the same organ, partly of a less obvious nature. In the latter case it may be assumed that a certain pattern of regulation is maintained by a line of cells descending from one mother cell. Other lines may have a slightly different pattern. To what extent heterochromatinization systems play a role here, is not known. Such variations, and the differences between sexes show that generalizations cannot always safely be made.

2.1.2 Crossing-over Estimated by Inference from Metaphase Association in Diploids (Disomics) and Autotetraploids (Tetrasomics)

2.1.2.1 The Relation between Metaphase Association and Crossing-over; Mapping Functions

Quite frequently, it is not possible to determine beyond doubt the exact number of chiasmata in a chromosome segment, for instance due to the fusion of chiasmata as a result of terminalization, or otherwise. Or the chromosomes may simply be too small, too condensed, or too fuzzy. Then there is no other possibility than to restrict the observations to scoring the presence or absence of one or more chiasmata: The criterion is mere association. Often, in material where chiasmata cannot be distinguished with sufficient certainty, early stages are not accessible and only metaphase is sufficiently clear to be studied. It is probable that in a number of cases of small chromosomes where merely the presence ore absence of chiasmata can be scored at metaphase, not more than a single chiasma may in fact be present (compare the small locust bivalents studied for instance by HEARNE and HUSKINS, 1935, and by HENDERSON, 1963). Then, scoring metaphase association is equivalent to scoring chiasmata. There is however seldom any certainty in this respect.

The analysis of crossing-over on the basis of metaphase association has some special limitations. As is the case with chiasmata, some metaphase associations may have been released before the time of observation. This can be generally detected when diakinesis data are compared with metaphase data: the latter then show lower values than the former. There is another serious drawback, which in a number of instances makes it impractical to work with the frequency of metaphase association for estimating crossing over: some chromosome segments (long arms, for instance) may almost without exception have at least one chiasma. An example is again the observation on the locust chromosomes: even the short chromosomes which in a sample of 180 cells never had more than one chiasma, also appeared never to have *less* than one. As long as it can be determined with certainty that there always is only one, this gives a good estimate of over-all crossing-over in this chromosome. However, when nothing more than metaphase association can be scored, and the segment considered is invariably associated, this cannot be meaningfully interpreted in terms of chiasma frequency. When in a significant proportion no association is observed, at least some information can be extracted, even when it is uncertain how many chiasmata are involved in the association.

Starting from the assumption (certainly not generally valid, as seen in 2.1.1) that within a chromosome the chiasmata are formed in a random fashion, their distribution should follow a Poisson series. This is also true for any segment of the chromosome. In fact, even when different segments of the same chromosome have a different probability of forming a chiasma (i.e. when chiasma formation is localized) for each segment separately the Poisson distribution is the "expected" distribution. Since Poisson distributions are additive, the distribution of entire chromosomes (or arms, or any segment) with zero, one, two, three etc. chiasma follows a Poisson series even with localized chiasma formation. The basic assumptions are that within each segment, however small, the formation of a chiasma is a random process, and that formation of a chiasma in one segment does not affect the probability of formation of a chiasma in another segment. Especially the second assumption is not correct, but this will be dealt with later.

The probability of a chromosome (or any other segment) having no chiasma at all now is $e^{-\mu}$, where μ equals the average chiasma frequency in that arm or segment. Since we are interested in estimates of crossing-over rather than of chiasma frequencies, it is useful to transform chiasma frequencies into crossing-over frequencies right at the beginning. Since every cross-over corresponds to two chiasmata, taking α as the average number of cross-overs in the segment considered, the frequency of the segment not being bound, i.e. having no chiasmata, equals $e^{-2\alpha}$. The frequency of the segment having at least one chiasma, i.e. being bound,

$$a = (1 - e^{-2\alpha}).$$

In 1919 HALDANE published a theoretical study on the relation between observed genetic *recombination* between two marker genes and the actual number of *cross-overs* expected to have been formed between them. With relatively short segments between the genes, when the number of cross-overs never or only very infrequently exceeds one, there is an almost complete correspondence between the frequency of recombination and of crossing-over. However, when the frequency of crossing-over increases and when in a number of cases two will occur in the same segment, the effects will cancel each other (Fig. 2.3) and no recombination is observed. When three occur in the same segment, there is recombination again. The increase in recombination levels off with increasing crossing-over. In fact, it can be shown that however high crossing-over becomes, recombination never exceeds 0.5 which equals that between genes on different chromosomes. These considerations led HALDANE to develop a function for the quantitative relation between recombination and crossing-over:

$$y = \tfrac{1}{2}(1 - e^{-2x})$$

in which y is the recombination frequency between two markers and x is the number of cross-overs. The main reason behind the interest in this relation is the fact that, whereas *numbers* of cross-overs are additive, recombination *frequencies* are not. When in genetic analyses the genes are placed on their chromosomes and the distances between them determined, it is customary to express these distances in cross-over frequencies. What is obtained is a linear (additive) *genetic map*. Although, as seen above, for short distances recombination frequencies suffice and can be equated to cross-over frequencies, this is not so for longer distances.

With the use of HALDANE's function a non-additive metric (recombination frequency) is transformed into an additive metric (number of cross-overs). It is, therefore, called a *mapping-function*.

The resemblance between HALDANE's mapping function and the function derived for the relation between the frequency of a segment being bound and the average number of chiasmata, is striking. It appears that the *probability of being bound* is *twice the recombination frequency* for any chromosome segment. This agrees with the fact that the maximum possible frequency of being bound equals unity, while the maximum recombination frequency equals 0.5 (cf. MATHER, 1936).

In HALDANE'S mapping function and in the formula relating the frequency of being bound to the average number of chiasmata, the distributions of cross-overs and of chiasmata respectively were assumed to be random. There is one major factor that causes these distributions to deviate considerably from randomness: *interference.* Interference implies that the formation of a chiasma restricts the probability of formation of another chiasma in the immediate neighborhood, i.e. the chiasmata tend to be separated by intervals of a certain minimum length. Thus in any segment of a chromosome the frequency of two or more chiasmata being formed is lower than expected on the basis of the average chiasma frequency, while the frequency of only a single chiasma being formed is higher. This means that the correction for double chiasmata is exaggerated in HALDANE'S mapping function, i.e. the map distance (= number of cross-overs) is not as high in reality as calculated from the recombination frequency using HALDANE'S function. Similarly, the map distance is exaggerated when calculated from the frequency of being bound, using the formula given above.

Several workers have attempted to construct mapping functions in which the effect of interference is included. Since interference is not predictable on purely theoretical grounds, the mapping function must contain an empirical component for interference. Nor is interference constant. A generally applicable mapping function therefore is based on average interference and can never be very accurate. Exact information on interference in any particular case is extracted only by detailed genetic or cytological analysis, which at the same time results in accurate maps. Thus mapping functions almost by definition are only rough approximations, as they are to be applied specifically in cases where nothing is a priori known about interference and only (relatively large) recombination frequencies are available.

In 1944 KOSAMBI published a partly theoretically derived function that comes close to accounting for interference as it is observed in the average *Drosophila melanogaster* chromosome:

$$y = \tfrac{1}{2}(th\ 2x)$$

in which the hyperbolic tangent (*th*) equals $\dfrac{e^{-2x} - e^{-2x}}{e^{-2x} + e^{-2x}}$. For distances longer than 35 map units, the function is not very satisfactory. OWEN'S (1950) function is somewhat better for longer distances, but of course also remains an approximation: $y = \tfrac{1}{2}(1 - e^{-2x}\cos 2x)$. CARTER and FALCONER'S (1951) mapping function is often used in human cytogenetics, and appears more favorable for most distances than KOSAMBI'S. It reads:

$$x = \tfrac{1}{4}(th^{-1} 2y + tg^{-1} 2y).$$

For short distances x and y are almost equal, for long distances where the recombination percentage approaches 50, no meaningful estimates of map distance are possible.

The mapping functions used for converting recombination frequencies into cross-over frequencies can equally well be used for converting frequencies of being bound into cross-over frequencies for specific chromosome segments. Their applicability, however, is limited by the circumstance that, especially when entire chromosome arms are studied, quite high frequencies of being bound are encountered, often approaching 1, usually not below 0.8, which corresponds with 40% recombination or a map length of 55 units on the basis of KOSAMBI'S function. This is a range in which the known mapping functions are not very accurate.

It is now necessary (and possible) to discuss the problem of arithmetically averaging frequencies of being bound. This is often done without further considerations, and is in fact inevitable when the bivalents (or bivalent arms) in a cell cannot be distinguished and have to be pooled. Each has its own particular probability of metaphase I association, but only the sum or average is determined for each cell. There is no satisfactory way out of this dilemma, but it is useful to give a few hypothetical examples which demonstrate the significance of the problem and which illustrate the magnitude of the error introduced. The *numbers* of chiasmata in the chromosomes or arms can be averaged without complications, and these numbers can be approximately derived, with the use of mapping functions, from the frequency of metaphase association. For instance, a chromosome (or arm) which is bound with a frequency of 0.6 has an average of 0.458 cross-overs on the basis of the function of HALDANE, and 0.345 cross-overs according to KOSAMBI'S function. This corresponds with 0.916 and 0.690 chiasmata respectively. Another chromosome (or arm) with a frequency of being bound of 0.9 has an average cross-over frequency of 1.151 (HALDANE) or 0.735 (KOSAMBI). This corresponds with 2.302 and 1.47 chiasmata respectively. The average of the two chromosomes for frequency of being bound is 0.750, but for number of chiasmata it is 1.609 (HALDANE) or 1.08 (KOSAMBI). Reconverting to (average) frequency of being bound gives 0.800 (HALDANE) or 0.793 (KOSAMBI), which deviate considerably from 0.750. When (for instance by insertion of a large piece of chromosome, or by special environmental or genetic conditions) the number of chiasmata is doubled in both, the numbers of chiasmata now are 4.604 or 2.94 respectively for the "long" chromosome, and 1.832 or 1.38 respectively for the "short" chromosome. The corresponding frequencies of being bound increase to 0.99 (HALDANE) or 0.994 (KOSAMBI) for the "long", and 0.84 (HALDANE) or 0.881 (KOSAMBI) for the "short" chromosome, averages 0.915 or 0.938.

This example gives an impression of the usefulness of transforming the frequencies of being bound into numbers of chiasmata, when manipulating observations. In this case the mapping functions serve merely as transformation functions, and even for this purpose the more sophisticated functions are preferred over HALDANE'S function, which considerably over-estimates the numbers of chiasmata, just as it neglects interference.

2.1.2.2 Frequencies of Being "Bound" in Diploids (Disomics) and Autotetraploids (Tetrasomics)

When a particular chromosome can be recognized at metaphase and has only one arm in which chiasmata are formed (telocentrics or pronounced acrocentrics), the frequency of association can readily be determined provided enough cells are available for analysis. With two arms considerably different in length, both can most probably be recognized and for each an estimate can be made directly. Examples are given in the section on interference (2.1.2.3). Quite frequently however, the two arms of a chromosome differ in length insufficiently to make a distinction at meiotic metaphase possible. Then a statistical approach may be tried. Say a (sub)metacentric bivalent with arms A and B (Fig. 2.9A) has an MI association frequency (the frequency of being "bound") a for arm A and b for arm B. When both arms are "bound" the bivalent has a ring shape. When arm A is bound but B is not, the bivalent has a rod shape. A similar rod-shaped bivalent appears when arm B is bound, but A is not. When neither A nor B are bound, two univalents appear. These three configuration types can by distinguished and their frequency recorded. The frequency r of ring-bivalents can be expressed in terms of the frequency of both A and B being bound or

$$r = a \cdot b. \tag{2.1}$$

The frequency of rod-shaped or open bivalents

$$o = a \cdot (1-b) + b \cdot (1-a). \tag{2.2}$$

The frequency of univalent pairs

$$u = (1-a) \cdot (1-b). \tag{2.3}$$

The sum of the three equals 1, and consequently there are two independent equations with two unknowns (a and b). These can be solved. The following example of rye is taken from SYBENGA (1965a). In a total of 100 cells there were 700 chromosome pairs; 509 were ring bivalents, 180 open bivalents and 11 univalent pairs. When these numbers are converted to fractions,

$$r = \tfrac{509}{700} = 0.726 = a \cdot b,$$

$$o = \tfrac{180}{700} = 0.257 = a + b - 2ab.$$

These two equations suffice. They are symmetric for a and b. By substitution the quadratic equation $x_{ab}^2 - 1.709\, x_{ab} + 0.726 = 0$ is derived, with roots

$$x_{a,b} = \frac{1.709 \pm \sqrt{1.709^2 - 4 \times 0.726}}{2} \quad \text{or,}\ a = 0.917\ \text{and}\ b = 0.790.$$

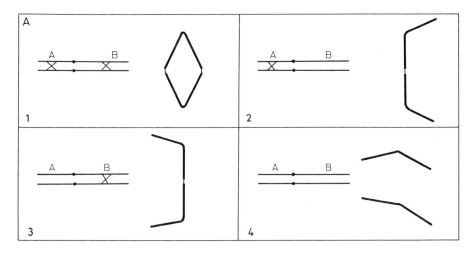

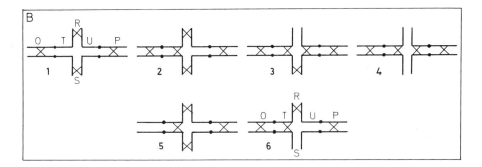

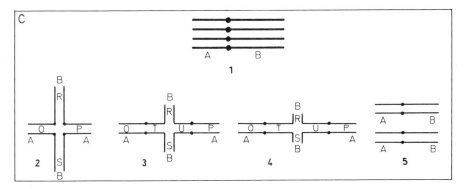

Fig. 2.9 A–C. M I association in relation to chiasma formation (arms *A* and *B* indistinguishable at meiosis). (A) Disomic. (B) Tetrasomic; all arms associated (bound) in *1, 2, 3* and *4*; one arm not bound in *5* and *6*. (C) The shifting point of partner exchange in tetrasomics with quadrivalent pairing (*2, 3, 4*); bivalent pairing in *5*

In this particular example all bivalents of all cells were pooled as there was no means of distinguishing between bivalents. The values obtained must therefore be considered not only average values of cells, but also of bivalents. Thus, *a* represents the "average" long arm and *b* the "average" short arm. This is not a strictly satisfactory procedure, as different chromosomes have different ratios of frequencies of being bound for the long arm and the short arm, as discussed in 2.1.2.1.

A similar analysis in an autotetraploid is more complicated. Each chromosome has four homologues, which implies the necessity of making a choice for pairing, since at any specific site only two chromosomes pair (Fig. 2.9 B). Occasionally, pairing initiates in only one point per chromosome and then only bivalents are formed. When a chromosome pairs with one partner at one site and with another at a second site, partner-exchange must have occurred between the two sites, and multiple associations *(multivalents)* result. In the simplest case partner exchange occurs at or near the centromeres, leaving entire arms for pairing and chiasma formation (Fig. 2.9 C 2). Partner exchange is usually not restricted to the centromeric regions and in principle may occur anywhere (Fig. 2.9 C 3,4) in the chromosome or it may even occur twice in the same chromosome. Then it is possible that within a single arm chiasmata are formed on both sides of the point of partner exchange, and entirely new types of configuration become possible. In such configurations the classification of a chromosome arm as being bound or not bound can become equivocal. Since such situations will be met every time configurations with partner exchange within an arm occur, it is worth while paying some attention to them.

When a point of partner exchange occurs anywhere within an arm, we can distinguish *end segments*, also called pairing segments (R, S in Fig. 2.9 C) and *interstitial segments* (T, U in Fig. 2.9 C). In both, or in either, chiasmata can be formed. There are also the arms O and P (homologous arms in the case of a tetrasomic or autotetraploid, but not experiencing partner exchange in this example). Since we are concerned with arms being bound, and not with chiasmata, arms with chiasmata on both sides of the point of partner exchange are not in any respect bound more than arms with chiasmata on only one side of the point of partner exchange. All configurations of Fig. 2.9 B 1, 2, 3, 4 have all arms bound, but no more. In Fig. 2.9 B 5, one arm pair is not bound, as is clear without further discussion. In Fig. 2.9 B 6, however, the same is true, although it appears that at least one of the two partners of the pair of arms with end S is bound (*viz.*, in T). Nevertheless, this arm is counted as entirely *not*-bound, for the following reasons: the entire arm consisting of segments U and S is devoid of chiasmata. There *must* be the equivalent of one other entire arm without chiasmata. This consists of two parts: the U segment of the arm composed of U and R, and the S segment of the arm composed of T and S. Thus there is a total of two arms not bound, one of which is represented by two segments in different chromosomes. The extra chiasma in T is, as it were, counted as an extra chiasma belonging to the arm pair associated at R. This method of counting one arm pair as having no chiasmata on basis of the fact that one of its partners lacks chiasmata may look somewhat arbitrary in an autotetraploid; it is more logical in the very similar configuration found in translocation heterozygotes (Section 2.2.4). In order to maintain the parallel with translocation heterozygotes, but also because it has advantages in autotetraploids with shifting points of partner exchange, this method has been adopted in the example on rye B-chromosome tetrasomics (SYBENGA and DE VRIES, 1972). Crossing-over in B-chromosomes in rye has been considered above for the disomic condition. The two arms can be distinguished, but within an arm there is no certainty about the number of chiasmata. The short arm may safely be assumed to have a maximum of one, but from the occurrence of MI configurations with partner exchange it is clear that in the long arm at least two may be formed with an appreciable frequency. In Fig. 2.10 the possible origin of a number of MI configurations from different combinations of chiasmata in the bivalent and quadrivalent pairing situations is shown. Since the number of theoretically possible configurations is high, only those actually observed are considered. It will not be difficult to determine the number of bound arms in configurations other than those shown here. Table 2.15 gives the configurations observed, a reference to Fig. 2.10 as far as their derivation is given there, the frequencies of being bound for the two arms in each configuration (0, 1 or 2), and the observed numbers and percentages for three plants. Some of the configurations shown may have a variable appearance with varying orientations of the centromeres, but this is of no consequence for chiasma studies.

It is clear that the chiasma frequency in the short arm is much lower than that in the long arm, and it is worth while comparing this difference with the physical arm-length ratio. Frequencies of being bound as such are not very useful for this purpose, but conversion to chiasma frequencies with any acceptable mapping function will do (see 2.1.2.1). Although not theoretically perfect for the intervals covered, KOSAMBI'S (1944) function can be used. For the three tetrasomics the cross-over frequencies $\alpha = 0.026, 0.041$ and 0.008 respectively can be derived for the short arm, and $\beta = 0.637, 0.686$ and 0.873 for the long arm. For a comparison, the disomic values are: $\alpha = 0.063$ and 0.059 for the short arm and $\beta = 0.628$ and 0.653 for the long arm. The tetrasomic arm-ratios are 12.3; 8.4 and 52.3; those for the disomics 5.0 and 5.6. The difference is striking and not readily explained. The physical (somatic) arm-length ratio in this material was 3.2 in squash preparations after bromonaphthalene pretreatment. Apparently the short arm forms chiasmata much less frequently than expected on the basis of its length. One reason may be that B-chromosomes have a sizable heterochromatic segment on each side of the centromere which has a function in its peculiar behavior in the first pollen mitosis, leading to regular non-disjunction. In the short arm, this takes up a relatively much larger segment than in the long arm.

Quite frequently in tetrasomics, the two arms cannot be distinguished as readily as in this example. This is the case for instance in tetraploid *Tradescantia virginiana*, where the chromosomes are metacentric or sub-metacentric. It has the additional disadvantage that the different chromosomes are also very similar at meiosis, and have to be pooled. An advantage of this plant is that partner exchange is predominantly at the centromere, and that, consequently, configurations with interstitial chiasmata are scarce. When they occur, they can be combined without much complication with those simpler configurations with which they correspond. Then the number of possible configuration types used for estimating the frequencies of being bound can remain small. The example is from SYBENGA

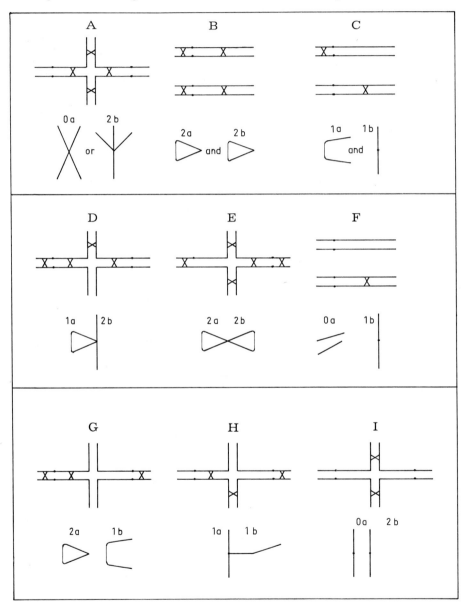

Fig. 2.10A–I. Nine of the many possible M I configurations of a tetrasomic with two distinctly different arms (B-chromosomes of rye). Types (B), (C) and (F) can also be derived from quadrivalent pairing; (G) and (I) also from bivalent pairing. The number (*a*) of short arms and (*b*) of long arms bound is given for each configuration. The two shapes of (A) are alternatives in respect to orientation, and are equivalent in respect to chiasma formation. (SYBENGA and DE VRIES, 1972)

(1975). Pairing is partly in bivalents and partly in multivalents (quadrivalents, or trivalents with univalents). It is again possible to determine how many chromosomes have both arms bound (Fig. 2.11A): two in a *ring bivalent*, four in a

Table 2.15. B-chromosome tetrasomics of rye: the 19 configuration types observed; *one* of the modes of origin for nine of these configurations as given in Fig. 2.10; the number of short (a) and long (b) arms bound in each of the configurations; their number and % frequency in three plants; and the average frequency of each arm being bound (SYBENGA and DE VRIES, 1972)

Configuration		Fig. 2.10	arms bound		plant 67277-9		plant 67383-11		plant 69385-2	
			short(a)	long(b)	number	%	number	%	number	%
X-Y	1	A	0	2	104	17.11	31	7.75	75	12.50
⊣	2	D	1	2	5	0.82	8	2.00	1	0.17
⊢⊢	3	H	1	1	4	0.66	8	2.00	—	—
⌊⌋	4	—	2	1	1	0.16	—	—	—	—
∪	5	—	1	2	21	3.45	7	1.75	2	0.33
⟨⟩	6	—	2	2	1	0.16	2	0.50	—	—
⋈	7	E	2	2	—	—	1	0.25	—	—
Y-	8	—	0	1	61	10.03	40	10.00	15	2.50
Y-	9	—	1	1	3	0.49	5	1.25	—	—
⋋	10	—	1	1	1	0.16	3	0.75	—	—
⊳ ⊳	11	B	2	2	—	—	1	0.25	1	0.17
⊳ |	12	—	1	2	19	3.13	17	4.25	12	2.00
⊳⊰	13	G	2	1	—	—	1	0.25	—	—
||	14	I	0	2	290	47.69	243	60.75	440	73.33
⊰|	15	C	1	1	1	0.16	6	1.50	2	0.33
⊰=	16	—	1	0	1	0.16	—	—	1	0.17
|=	17	F	0	1	84	13.82	25	6.25	50	8.33
⊰=	18	—	1	1	4	0.66	1	0.25	—	—
==	19	—	0	0	8	1.32	1	0.25	1	0.17
					608	99.99	400	100.00	600	100.00

	plant 67277-9	plant 67383-11	plant 69385-2
Average arms bound *a*	0.0518	0.0813	0.0167
Average arms bound *b*	0.8544	0.8788	0.9409
Arms bound rest genome	0.976	0.971	0.821

ring quadrivalent, two in a *chain quadrivalent*, one in a *trivalent*. There are also chromosomes with one arm bound: two in an *open bivalent*, two in a *chain quadrivalent*, two in a *trivalent*. Chromosomes without any arm bound are *univalents*. This analysis is best carried out with a single chromosome (arm) as a unit instead of an (arm) pair, as has routinely been possible in the diploid, but this is of no consequence. In the example mentioned, 25 cells were studied, each with 24 chromosomes, or a total of 600 chromosomes. There were 31 ring quadrivalents, 43 chain quadrivalents, 4 trivalents, 92 ring bivalents, 48 open bivalents and 12 univalents. Thus the frequency of chromosomes with both arms bound $(a \cdot b)$ equals $\dfrac{4 \times 31 + 2 \times 43 + 1 \times 4 + 2 \times 92}{600} = 0.663$. Chromosomes with one arm bound and not the other: $a(1-b)+b(1-a)=a+b-2ab$ $=\dfrac{2 \times 43 + 2 \times 4 + 2 \times 48}{600} = 0.317$. The remainder has no arms bound or $(1-a) \cdot$

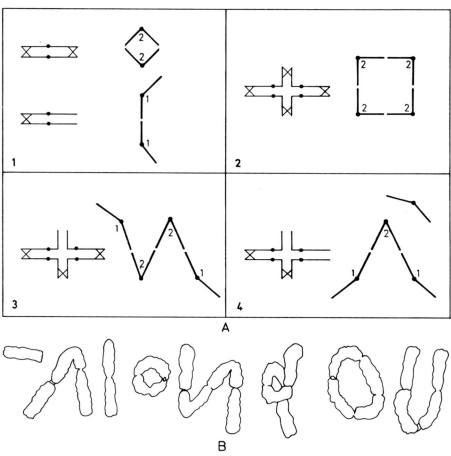

Fig. 2.11A and B. Bivalent and multivalent formation in a tetrasomic with metacentric chromosomes. (A) The derivation of some of the M I configurations; for each chromosome the number of arms bound is shown at the centromere. (B) The same configurations as observed in autotetraploid *Tradescantia virginiana*. (SYBENGA, 1975)

$(1-b) = \frac{12}{600} = 0.020$. The total, of course, equals 1. The corresponding quadratic

equation is $x^2 - 1.643x + 0.663 = 0$ and the roots are $x_{a,b} = \dfrac{1.643 \pm \sqrt{1.643^2 - 2.652}}{2}$,

and $a = 0.931$; $b = 0.713$.

More complicated diakinesis and metaphase configurations with one point of partner exchange in autotetraploids closely resemble those of translocation heterozygotes (2.2.4).

2.1.2.3 Interference

Without direct observation of chiasmata, the detection and estimation of interference is limited and much less exact than when the frequency and location of chiasmata can directly be determined. When at least two different segments can be distinguished it is still relatively simple to ascertain whether chiasma formation in these segments is independent or not. With independence, the frequency of simultaneous occurrence of chiasmata in both segments must equal the product of the frequencies for the segments separately.

In unmarked chromosomes, the only segments available are the two arms, and detection and estimation of interference is possible only when in both arms absence as well as presence of chiasmata are sufficiently frequent. When one always (or never) has at least one chiasma, there are no means of analyzing interference.

With a frequency a for the shorter of the two arms being bound, and a frequency b for the longer arm, the expected combined association frequency is $a \cdot b$. Only the short arm $a \cdot (1-b)$, only the long arm $(1-a) \cdot b$ etc. In Table 2.16 an example is given for a B-chromosome pair of rye (SYBENGA and DE

Table 2.16. B-chromosome disomics of rye: bound arm frequencies, coincidence and interference (SYBENGA and DE VRIES, 1972)

	Plant 67368-9		Plant 69382	
	Cells	Frequency	Cells	Frequency
Both arms bound	57	0.095	64	0.107
Short arms only	18	0.030	6	0.010
Long arm only	453	0.755	455	0.758
No arms bound	72	0.120	75	0.125
	600	1.000	600	1.000
Non-B bivalents, average		0.984		0.983
Short arm bound (B)		0.125		0.117
Long arm bound (B)		0.850		0.865
Both expected to be bound	63.6	0.106	60.6	0.101
Contingency χ^2 from a 2×2 table of the observed numbers	5.445 P=0.02		1.648 n.s.	
Coincidence	$\dfrac{57}{63.6} = 0.898$		$\dfrac{64}{60.6} = 1.056$	
Interference	$1 - 0.898 = 0.102$		$1 - 1.056 = -0.056$	

VRIES, 1972). The computed random frequencies were converted to numbers and compared with the observed numbers for the χ^2 test. Using a similar procedure as NEWCOMBE (1941) for Trillium and CALLAN and MONTALENTI (1947) for Culex, interference was computed as (1-coincidence), the latter being the ratio between observed and expected frequencies of simultaneous association of the two arms. PÄTAU (1941) developed a different method, inspired by HALDANE'S (1931) approach, which used the ratio between the observed variance and the expected variance (= the mean) as an index of interference. When the observed variance is smaller, there is positive interference. Using his method on the small, subacrocentric chromosome of *Dicranomyia trinotata*, PÄTAU (1941) could conclude that there was strong and consistent positive interference across the centromere, in spite of the fact that different animals differed greatly in respect to the frequency with which the two arms were bound by chiasmata. This variation was ascribed to chromosomal rearrangements, floating in the population, but apparently not affecting across-centromere interference. When working with bound arms, and not chiasmata, this procedure is not especially complicated, but the first method is somewhat simpler and more lucid.

In the example of the B-chromosomes of rye, one of the two plants showed significant across-centromere interference, the other not. Such a difference may result from genetically conditioned differences in chiasma localization in the long arm (cf. 2.1.1.2 and CALLAN and MONTALENTI, 1947). The more proximally located, the more a chiasma in the long arm will have an effect on the short arm. Such localization could not be studied in the disomic, but an analysis of tetrasomics gave very indirect indications in this respect.

The analysis of interference is much more complicated in tetrasomics than in disomics. The set of four chromosomes can be associated as a quadrivalent, a trivalent with univalent, two bivalents (rings and/or open bivalents), a bivalent with two univalents, or four univalents. The simplest way for analyzing over-all interference (it can be analyzed separately for the bivalent configurations) is first estimating a and b and then calculating the expected frequencies of two bound short arms, a^2; one bound short arm, $2a \cdot (1-a)$; no bound short arm, $(1-a)^2$. The same is done for the long arm (b). From these expectations the nine combinations are constructed and compared with the frequencies of configurations in which these specific combinations occur. These can be derived from Table 2.15 for B-chromosome tetrasomics in rye (SYBENGA and DE VRIES, 1972).

The expected and observed combinations for two tetrasomic plants are given in Table 2.17. One plant shows a significant deviation: neglecting the low-frequency classes, it is seen that when both long arms are bound, b^2, there is a deficit of short bound arms, $2a \cdot (1-a)$, and an excess of short not-bound arms $(1-a)^2$. When the number of bound long arms decreases, the number of short bound arms increases: positive interference across the centromere. In the other plant expected and observed frequencies correspond perfectly. The plant with interference had the highest frequency of short arms bound, but in the non-multivalent configurations this frequency was much lower than in the multivalent configurations. It is probable that in this particular plant partner exchange took place near the end of the long arm, leaving ample opportunity for interstitial chiasmata, but reducing the chance for multivalent formation. This is a consequence of suppression of the terminal chiasma by distal partner exchange. Interstitial chiasmata are rather close to the centromere, and thus may have a considerable effect on the short arm.

Table 2.17. B-chromosome tetrasomics of rye: relation between chiasma formation in long and short arms. For the average frequencies of short arms bound (a) and long arms bound (b) see Table 2.15. In each sub-table the nine combinations of: both short arms bound (a^2), one short arm bound, the other not ($2a(1-a)$) and no short arms bound ($(1-a)^2$), with the same three possibilities for the long arm, b. In each block the expected (independence) frequency, the expected number and the observed number. Derivation of observed numbers from Table 2.15, columns 6, 8 and 10 by adding the frequencies of configurations having the same combinations of arms bound or not bound. Two plants. For plant 67383-11, $\chi_3^2 = 17.27$; P<0.01 (SYBENGA and DE VRIES, 1972)

Plant 67277-9

b \ a	a^2	$2a(1-a)$	$(1-a)^2$
b^2	0.0020 1.22 1	0.0717 43.59 45	0.6563 399.03 394
$2b(1-b)$	0.0007 0.43 1	0.0244 14.84 13	0.2237 136.01 145
$(1-b)^2$	0.0001 0.06 0	0.0021 1.28 1	0.0191 11.61 8
		Total	1.0001 608.07 608

Plant 67383-11

b \ a	a^2	$2a(1-a)$	$(1-a)^2$
b^2	0.0051 2.04 4	0.1152 46.08 32	0.6518 260.72 274
$2b(1-b)$	0.0014 0.56 1	0.318 12.72 23	0.1798 71.92 65
$(1-b)^2$	0.0001 0.04 0	0.0022 0.88 0	0.0124 4.96 1
		Total	0.9998 399.92 400

In numerous cases the two arms of a bivalent cannot be distinguished, and then the possibilities become even more limited. One can accept the presumption of PÄTAU (1941) and of CALLAN and MONTALENTI (1947) (cf. 2.1.1.2), that the two arms have the same chiasma frequency, but this is in most cases not realistic. Then there is one approach left, which is of some general significance, and will be shown to be of special importance in translocation heterozygotes. It only detects negative interference, which is not very common, but because of the general significance of the analysis and the insight it gives in some of the problems encountered with bivalents with almost equal arms, it will be treated in some detail. It is a mathematical approach relating to the conditions that must be satisfied for the solution of quadratic equations such as Eq. (2.4). The discriminant in this equation is $(2r+o)^2 - 4r = 4r(r+o-1) + o^2 = o^2 - 4ru$. For the roots to be solvable the discriminant must be positive. It is so only when four times the product of the frequencies of ring bivalents (r) and univalents (*u*) does not exceed the square of the frequency of the open bivalents (*o*). This means that, when the combined frequency of rings and univalents exceeds a certain value, there are no roots, i.e. that the assumptions on which the equations are based apparently are not correct. The major assumption for these equations is that the probabilities of formation of chiasmata in the two arms may be multiplied, i.e. that the events involved are independent. Therefore, a negative discriminant is an indication that the events, i.e. chiasma formation in the two arms, are *not* independent, which of course means that there is interference. Since in this case the simultaneous occurrence and non-occurrence of chiasmata are in excess (rings and univalents), it is *negative interference*. Negative interference is not detected equally well for different ratios of genetic length of two chromosome segments. The theoretically optimal situation is that in which the two arms are equal: the discriminant should equal 0, and when negative, this is entirely due to negative interference. In such cases, however, sampling error will cause the discriminant to be negative in about half the cases and to be positive in the other half. Statistical analysis must then decide if the effect is significant.

In fact, the two independent equations on which the estimates of the genetic arm lengths are based (any two out of Eqs. 2.1; 2.2; 2.3) do not permit estimating a third parameter (their interaction = interference). Only in the marginal case where the equations do not yield a solution (negative discriminant) is the detection of negative interference possible. Its intensity *beyond* the point where it compensates the difference in arm length can then also be estimated.

As long as the discriminant is positive, neither negative nor positive interference can be estimated or even detected. When there is negative interference, it will cause the estimated length of the two segments to appear more equal than they in reality are, but there is no means of detecting this. Similarly, when there is positive interference, the estimated genetic lengths of the arms will appear more different than they are, but this again remains undetected. The mutual dependence of the estimates of interferences across the centromere and of the difference between the two arms has been quantitatively formulated by PÄTAU (1941). Since he uses the ratio of the two frequencies of being bound, the value of this function and graph is limited for high frequencies. Interference clearly is of great consequence for the estimation of genetic length by the methods described here, even when for the actual conversion of frequency of being bound into map length, the proper mapping function is employed. It is usually assumed that there is no interference between the two arms of one chromosome, although there are many exceptions (see 2.1.2.2 and above). For the analysis of adjacent segments in the same chromosome arm, as in translocation heterozygotes (2.2.4), interference can be even more disturbing.

In this Section the two basic aspects of estimating cross-over frequencies from metaphase configuration frequencies, without observing chiasmata, have been considered: (1) Estimating the frequencies of being bound from the relative frequencies of different configurations originating from the same pairing configuration and (2) Deriving the cross-over frequencies from the frequencies of being bound with the use of mapping functions.

On these two principles much of the remainder of this Chapter will be based.

2.2 Marked Chromosomes: Structural Deviants

2.2.1 Telocentric Substitutions

Chromosome-structural deviations which lead to specific meiotic configurations can be used to mark chromosomes which otherwise are undistinguishable. The rearrangement with the least genetic and cytological complications is the substitution of a metacentric chromosome by the two corresponding telocentrics. It is a rather uncommon type of deviation, which is not readily established. Artificial telocentric substitutions are available in wheat, in barley, and in rye. Of the latter, quantitative analyses have been made of the relative frequencies of the different configurations. In a number of genera of plants and animals, different species have a different chromosomal constitution as a result of Robertsonian rearrangements: some have certain chromosomes as metacentrics, other species have the same chromosomes represented by the two corresponding telocentrics. In hybrids between such species, a heterozygous telocentric substitution is present. Well known examples are the tobacco mouse *(Mus poschiavinus)* vs. the common house mouse *(Mus musculus)*. In several lizard and insect species similar Robertsonian rearrangements have occurred.

Both homozygous and heterozygous telocentric substitutions can be used, but only when not more than a single metacentric chromosome is involved is the substitution to be used as a marker for a specific chromosome.

Two situations are to be distinguished: (a) the two arms are morphologically distinct; (b) the arms are not morphologically distinct. In the first case it is possible to make direct observations on each arm and to detect and estimate possible interference. In the second case equations very similar to those of (2.1), (2.2) and (2.3) are used to determine the frequency of each arm being bound, and only detection of extreme negative interference is possible. Fig. 2.12 shows the chromosomes of a heterozygous telocentric substitution and the meiotic configurations corresponding with specific combinations of chiasmata. It is readily seen that a trivalent results when both arms have at least one chiasma (frequency $a \cdot b$). A heteromorphic bivalent results from chiasma(ta) in one arm and none in the other (frequency $(1-a)b + (1-b)a$) and with no chiasmata at all, there are three univalents: $(1-a)(1-b)$. With homozygous substitutions with distinct arms again one simply counts the number of bivalents and univalents for each arm. When the arms are not distinct the frequency of two bivalents is $a \cdot b$, that of one bivalent and a pair of telocentric univalents is $(1-a)b + (1-b)a$ and that of four univalents is $(1-a)(1-b)$. In an example of a heterozygous telocentric substitution for the satellite chromosome of rye in which the two arms could not be recognized with sufficient certainty (in other material of the same substitution in rye the two telos can sometimes readily be distinguished) the following configuration frequencies were observed.

In 500 cells, 412 had a trivalent (0.824); 86 (0.172) had a bivalent and an univalent, and in two cells (0.004) there were three univalents. This gives for one arm a frequency of being bound of 0.974 and for the other 0.846. Since the mitotic arm-length ratio of this chromosome is approximately 1.25, it is interesting to determine the ratio of crossing-over between the two arms.

The ratio of being bound, of course, is useless for this purpose, but a transformation to map length makes a more meaningful comparison possible. With HALDANE'S function the map lengths are 182.5 and 93.5 respectively, a ratio of 1.95. With KOSAMBI'S function the two values are 108 and 62, ratio 1.74. Apparently there is a discrepancy between mitotic arm-length ratio and genetic arm-length ratio. One possible reason is that the short, nucleolus carrying arm has pairing difficulties around the nucleolus, leading to reduced crossing-over. A second possibility is that there is positive interference increasing the apparent difference between the arms in a substitution configuration. It is not probable that this has a serious net effect on arms as long as these. A third possibility

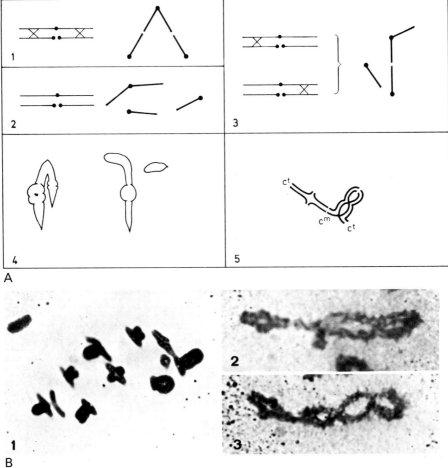

Fig. 2.12A and B. M I configurations and their derivation for a heterozygous telocentric substitution. (A) *1, 2, 3*: diagram; *4*: Camera lucida drawing of the M I configurations of diagrams *1* and *3* as observed in rye; *5*: interpretation of the trivalent of Fig. 2.12 (B) (*1*) with the centromeres marked: c^m for the metacentric, c^t for the telocentrics (in the short arm one distal chiasma, in the long arm three chiasmata). (B) M I (*1*) and diplotene (*2, 3*) configurations of a telocentric substitution heterozygote of *Acrida lata* (SANNOMIYA, 1968), cf. A 5

is that the sampling error is so great that not much value can be attached to the final results. In fact, there were only two cells in 500 with no chiasmata in both arms, and this is one of the critical categories. There are two arguments against this: first, in order to obtain an arm-length ratio (in this range of chiasma frequencies) close to 1.25, there should have been many more cells without chiasmata in this chromosome than is compatible with the frequency of 0.004 found. Secondly, in related material, in 1,200 cells not a single cell without chiasmata in this chromosome was encountered. This made an analysis actually impossible. It is probable, therefore, that the discrepancy between physical and genetic arm-length ratios is real.

The example given also illustrates the warning given at the beginning of this section: in some chromosome arms the chiasma frequency is too high to be estimated merely by means of the frequency of being bound: the number of configurations without chiasmata in this arm becomes critically low.

As noted above, natural centric fusion is equally efficient for marking a pair of otherwise not recognizable chromosomes. An example by SANNOMIYA (1968) for the insect *Acrida lata* is given in Table 2.18, cf. Fig. 2.12B.

Table 2.18. The frequencies of chiasmata in the long and the short arm, at diplotene and at metaphase I, of the heteromorphic trivalent of a centric-fusion heterozygote of *Acrida lata*, comparable with a telocentric substitution heterozygote (SANNOMIYA, 1968)

| Chiasmata | Diplotene | | M I | |
	Trivalents	%	Trivalents	%
Long arm				
2	45	26.47	83	28.72
3	116	68.24	200	69.20
4	9	5.29	6	2.07
Total	170	100.00	289	100.00
Average/trivalent		2.79		2.73
Short arm				
1	168	98.82	284	98.27
2	2	1.18	5	1.73
Total				
Average/trivalent		1.01		1.02

Although telocentric substitutions resemble the original situation very closely, the distribution of chiasmata may be slightly modified. This is suggested by genetic analyses of mono-telocentrics in cotton and wheat in which only one of the two telocentrics was present. ENDRIZZI and KOHEL (1966) reported that two genes in cotton, one on each side of the centromere, had a recombination frequency of 22.1% in normal chromosomes. In the respective mono-telocentrics, however, their respective distances from the centromere could be estimated as 1.0% and 4.4%, together 5.4%. A similar four-fold decrease was found by SEARS (1972b) in wheat, who observed a recombination frequency between two

markers each on one side of the centromere of 3.5%, but only a frequency
of 0.87% and 0.0% respectively between marker and centromere when studied
in the respective mono-telocentrics. For the entire arm the decrease seems to
be compensated by an increase more distally. In SEARS' (1972) report a recombination percentage of 43 is given for two genes in a telocentric, one close to
the centromere, the other rather distally located. In the normal chromosome
the percentage was identical (43.3). Thus when chiasma frequency in an entire
arm is considered, neglecting the location of the chiasmata, the effect may be
minimal for a relatively long arm.

2.2.2 Deficiencies

Deficiencies large enough do be recognized at meiosis are quite scarce. Yet
a few are known, even in diploids, and they can serve to mark specific chromosomes
and even specific arms. Usually the deficiency is available only in the heterozygous
condition, which does not make it less useful as a marker. The arm with the
deficiency may not be considered normal, and any information obtained in
respect to this arm only refers to the specific condition of the deficiency. The
other, non-deficiency arm of the same chromosome (if present), which in all
respects is normal, is also recognized: a deficiency is as effective a marker for
the non-deficiency chromosome arm as for the deficiency arm. Only when across-
centromere interference operates or when the deficiency affects the meiotic behavior of the entire cell, is the behavior of the other arm of the deficiency chromosome
to be considered abnormal. For an example a large radiation-induced terminal
deficiency in chromosome I of *Lilium formosanum*, reported by BROWN and
ZOHARY (1955) (Fig. 2.13) is used. Data are available on the frequency of chiasmata

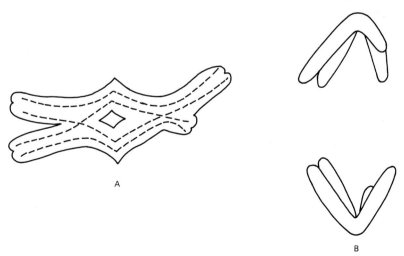

A

B

Fig. 2.13A and B. M I and A I configuration of a deficiency heterozygote with a chiasma
between the centromere and the deficiency, and one in the other arm. The deficiency
makes both arms recognizable at M I in a normally metacentric chromosome and reveals
exchange in the deficiency arm at A I

at diakinesis and the frequency of chromatid exchange observed at first anaphase. For BROWN and ZOHARY the purpose of the analysis was to demonstrate the correspondence between observed diakinesis chiasma frequencies in the deficient arm and observable exchange at anaphase, and they studied several deficiences and inversions. For the present purpose the data serve to show how to estimate crossing-over in the two arms of the marked bivalent when the deficiency is heterozygous. In this particular case direct observation of individual chiasmata at diakinesis was possible but when only metaphase association can be scored the approach is similar. Crossing-over in the deficiency arm could quite well be estimated at anaphase: it was so much shorter than its homologue that chromatid exchange between centromere and deficiency could be scored. In 155 completely analyzable cells at diakinesis 121 had one chiasma in the deficiency arm only; 10 had one only in the other, not marked but now recognizable arm. In 21 cells there was a chiasma in both arms, and in 3 cells neither arm had one. Thus $\frac{121+21}{155} = 0.902$ marked chromosome I bivalents had one chiasma in the deficient arm (between centromere and deficiency). This corresponds with 45.1 map units. In the normal arm at the other side of the centromere the frequency of chiasmata was $\frac{10+21}{155} = 0.200$, or 10 map units. In both arms simultaneously $0.2 \times 0.902 = 0.1804$ or 27.96 in 155 cells were expected, whereas 21 were observed, a slight and insignificant shortage, which might point to some (positive) interference across the centromere ($\chi_1^2 = 2.14$). At anaphase I the frequency of cells with a chromatid exchange in the deficient arm was 0.90. Apparently the agreement between the diakinesis and anaphase data is quite close. These observations were made in 1952. In the next year, 1953, the corresponding values were 0.79 at diakinesis and 0.78 at anaphase for the deficient arm. Since the same clone was used, the difference must have been environmentally conditioned. In another clone, a deficiency in a different chromosome was analyzed in 1951 both at metaphase and at anaphase. The non-deficiency arm had chiasmata in 100% of the cells studied, the deficiency arm had 0.7 chiasmata (35 map units) at metaphase and 0.71 cells with an observable exchange at anaphase I. A progeny plant had only 0.51 chiasmata in the deficient arm at metaphase and 0.55 exchanges at anaphase I. The difference is large and may be due to genetic and environmental effects.

In a comparable analysis of a deficiency heterozygote in *Allium fistulosum* by ZEN (1961), the deficient arm had a chiasma in 80.4% of the 769 cells observed at metaphase I and 79.5% chromatid exchange in 298 AI and AII cells.

A few more reports on deficiencies studied by other authors in different material are available, but these examples will suffice. A situation comparable, yet with fundamental differences is reported by SANNOMIYA (1968): in *Acrida lata* chiasmata could be counted in a heteromorphic bivalent, which was marked by the presence in one and the absence in the other homologue of a heterochromatic segment (Table 2.19). Such heteromorphism has been reported for more grasshopper and locust species, but in those cases recognition of chromosomes is usually possible also on the basis of other chromosomal characteristics.

Table 2.19. Chiasma frequencies at diplotene and M I in a (telocentric) heteromorphic bivalent of *Acrida lata* (SANNOMIYA, 1968)

Chiasmata	Diplotene		M I	
	Bivalents	%	Bivalents	%
1	15	21.43	68	23.45
2	54	77.14	222	76.55
3	1	1.43	—	—
Total	70	100.00	290	100.00
Average/biv.		1.80		1.77
Coincidence = ratio $\dfrac{\text{variance}}{\text{mean}^{a}}$		$\dfrac{0.191}{1.80} = 0.106$		$\dfrac{0.180}{1.77} = 0.102$
Interference		0.894		0.898

[a] Mean = expected variance.

2.2.3 Inversions

2.2.3.1 Paracentric Inversions

In paracentric inversions the centromere is outside the inverted segment: the inversion is confined to one arm. It is the only type possible in telocentric chromosomes. With complete pairing between a normal and an inversion chromosome, the inversion segment pairs in a loop (Figs. 2.14; 2.15). The *interstitial* (proximal) segment between centromere and inversion and the *distal* segment pair in a normal way. Usually, diakinesis and metaphase configurations are not particularly revealing, but a few analyses have been reported.

2.2.3.1.1 Diplotene-Diakinesis

At diplotene the inversion bivalent opens up like any normal bivalent. When there is no chiasma in the inversion segment, no detectable alteration in bivalent shape can be observed: the loop is stretched out, and nothing visible is left. An "inversion chiasma" (DARLINGTON, 1937, 1965) has an unusual appearance: instead of the normal structure the cross-over chromatids form two loops ("reversed chiasma" according to BROWN and ZOHARY, 1955) which can be recognized in favorable material. Origin and appearance are explained in Fig. 2.14. Especially in the "$" orientation the cross-over loops can be distinguished from cross-over chromatids in normal bivalents. In the "B" orientation this is usually more difficult. Whether a "$" or a "B" configuration is formed probably depends mainly on orientation in respect to the plane of squashing when a squash preparation is made, and therefore both types are expected to have equal frequency. Occurrence or absence of chiasmata in the distal segments determine the final shape of the bivalent, but they do not affect the character of the structure of the reversed chiasma. When the loops interlock (expected in 50% of the cases) it is very hard to distinguish a reversed chiasma from a normal one. The size of the loops is determined by the length of the inversion segment and the location of the inversion in the chromosome; further, when these exist, by the location of the external chiasmata. The loop sizes are not determined by the location of the chiasma inside the inversion, but this location does have an effect on the symmetry of the diplotene, diakinesis and metaphase configuration (Fig. 2.14A–B, *vs.* C–D). Consequently this is a means of distinguishing between chiasmata formed in the central or lateral parts of the inversion. Without external chiasmata, the asymmetry is expressed as a difference in length between the parallel arms on either side of the "chiasma". With external chiasmata, by a difference in length of the arm segments between inversion- and external chiasmata. Reversed chiasmata are recognized only in material with reasonably accurately traceable chromatids. With the B-type, recognition is often impossible even then, but the fact that at the "chiasma" only two of the four chromatids are continuous, and therefore stretched usually stronger than the four chromatids in normal chiasmata, helps in cases of doubt.

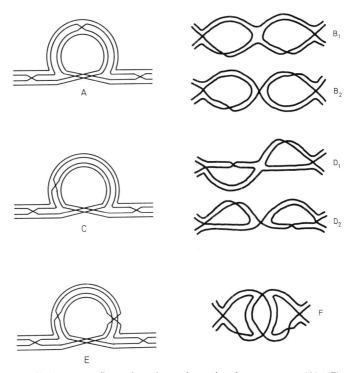

Fig. 2.14A–F. Pairing and diplotene configurations in an inversion heterozygote. (A), (C), (E) Pairing diagrams with different combinations of chiasmata. (B), (D), (F) The appearance of the "reversed chiasmata" in the corresponding diplotene configurations. (A), (B) Chiasma central in pairing loop. (C), (D) Chiasma lateral in pairing loop; (E), (F) two complementary chiasmata. The twist in the bivalents of (B_2) and (D_2) makes the recognition of the reversed chiasmata difficult in normal squash preparations

With two chiasmata in the inversion (Fig. 2.14 E, F), the compactness of the configuration is increased, and recognition is impeded. The B-type now is practically the only type possible, but it does not suffer from the drawbacks of the single-chiasma configuration. Interlocking of loops is now possible. When the two chiasmata are not located very close together (and this is not expected in most situations where interference operates) reciprocal chiasmata will not have an opportunity to fuse and remain visible as two separate chiasmata. Again, their relative position in respect to the end points of the inversion will determine the shape of the configuration (symmetric or asymmetric) and thus again the degree of symmetry is an indication of this location. Table 2.20 is based on observations reported by BROWN and ZOHARY (1955) on a paracentric inversion in *Lilium formosanum*. The frequency of bivalents with 1 and with 2 chiasmata in the inversion can be used to detect interference: comparison with a Poisson distribution based on the average indeed demonstrates a lack of bivalents with 2 chiasmata in the inversion. Interference is also indicated by the location of the chiasmata: with only one chiasma, 66 were found to be centrally located, 56 lateral. With two chiasmata in the inversion, 50 bivalents had both chiasmata lateral, 80 had one lateral and one central: in this group there were a total of 180 lateral chiasmata. Comparing the ratio of lateral to central chiasmata gives a significant difference: 56:66 when only one chiasma was present against 180:80 when there were two. As a total, there were 236 lateral and 144 central chiasmata. With random formation there would be a 2:1 ratio: apparently there is a shortage of lateral chiasmata, which may reflect pairing difficulties near the inversion ends or some general preference for formation in the central part of the inversion segment. The distinction between symmetric and asymmetric was generally clear, pointing to distinctly localized formation.

BROWN and ZOHARY (1955) also reported on the interaction between the inversion segment and the proximal and distal segments. Of 252 inversion bivalents with chiasmata in the inverted region, 251 had a chiasma in both the proximal and the distal segments, only one lacked a chiasma in the distal segment, but had one proximally.With no chiasmata in the inverted region (26 bivalents), the frequency of external regions (either proximal or distal) without chiasmata was also high: 7 bivalents. On the other hand, this was the only group in which in addition to a single chiasma in one of the external regions, two were formed in the other (3 bivalents). The latter points to positive interference, the former may be due to a general failure of pairing or of chiasma formation. If due to negative interference, then the combination of more than one chiasma in the inverted and in the external segments should also have been observed, but it was not.

2.2.3.1.2 Anaphase

A reversed chiasma in the inversion results in a bridge at first anaphase (Fig. 2.15A) accompanied by an acentric fragment. Thus when in an inversion heterozygote an anaphase bridge with a fragment is observed, there must have been a chiasma in the inversion segment. There are complications: although chiasmata in the distal segment have no effect on anaphase bridge formation, chiasmata in the interstitial segment can greatly influence the fate of the chromatid bridge formed by a chiasma in the inversion segment. When two different chromatids are involved in the interstitial and in the loop chiasma (complementary chiasmata),

Table 2.20 A and B. Chiasmata at diakinesis in two plants of *Lilium formosanum* heterozygous for a paracentric inversion (BROWN and ZOHARY, 1955)
(A) Chiasmata in inversion

Two chiasmata	Plant 221-11	Plant 221-19	Total
Symmetric	25	25	50
Asymmetric	36	44	80
Total two chiasmata	61	69	130
One chiasma			
Symmetric			
open	13	18	31
not distinguishable (interlocked)	19	16	35
Asymmetric			
B type	18	11	29
$ type			
open	7	6	13
not distinguishable (interlocked)	10	4	14
Total one chiasma	67	55	122
1 or 2 chiasmata subtotal	128	124	252
No chiasma	17	9	26
Grand total	145	133	278
Non analysable	9	8	15

(B) Chiasmata outside inversion in same arm

	Plant 221-11	Plant 221-19	Total
With chiasmata in inversion			
1 proximal, 1 distal	127	124	251
1 proximal, 0 distal	1	0	1
Total	128	124	252
No chiasmata in inversion			
1 proximal, 1 distal	9	7	16
2 proximal, 1 distal	0	1	1
1 proximal, 2 distal	2	0	2
1, proximal or distal	6	1	7
Total	17	9	26
Grand total	145	133	278

no effect results. The same is the case when the same two chromatids are involved (reciprocal chiasmata). However, when one of the two chromatids involved in one chiasma is also involved in the other, the bridge is transformed into a loop at anaphase I (Figs. 2.15A; 2.16). There is also a fragment. The loop may not always be recognized, but it turns into a chromatid bridge at second anaphase and this bridge may be scored. Anaphase II bridges, therefore, can be used for estimating crossing-over in the interstitial segment. The conversion of an anaphase I bridge into a loop is expected in 50% of the cases where a chiasma is formed in both the interstitial and the inversion segment, but when no chiasmata are formed in the inversion, chiasmata in the proximal segment go undetected. Two chiasmata in the inversion and one in the proximal segment may lead to the presence of a double bridge with fragments, or to

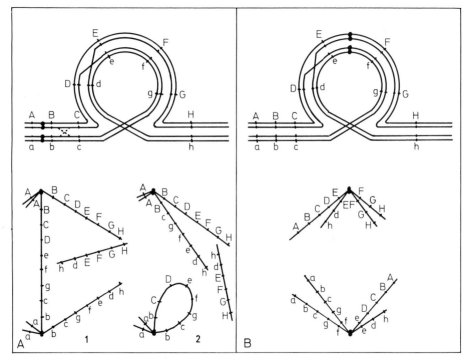

Fig. 2.15A and B. The anaphase configurations resulting from crossing-over in an inversion pairing loop. (A) Paracentric inversion; in 2 there has been a second (disparate) chiasma in the interstitial segment. (B) Pericentric inversion

a double loop with fragments, or to a single bridge with fragment, or to a single loop with fragment, or the two chiasmata may be reciprocal, cancel each other and become undetectable (Fig. 2.16). The relation between the frequencies of the different anaphase I configurations and the formation of chiasmata in the proximal (interstitial) and inversion segments were worked out by BRANDHAM

(1969) who distinguished six categories on the basis of the number of chiasmata (0, 1 or 2) in the inversion and proximal segments. These categories are given in Table 2.21, in addition to a seventh category added for completeness. Such

Table 2.21. The combinations of 0, 1 or 2 chiasmata in the inversion and in the interstitial segment of a heterozygous paracentric inversion, and the expected frequencies of normal anaphase I (N); a bridge and a fragment (BF); a loop and a fragment (LF); a double bridge and two fragments (BBFF); a double loop and two fragments (LLFF) for each combination (BRANDHAM, 1969). In addition, the observed frequencies in 1,500 cells of an inversion heterozygote in *Agave stricta* (BRANDHAM, 1969), and in 318 cells of *Lilium formosanum* (BROWN and ZOHARY, 1955)

| Category | Number of chiasmata | | Anaphase configuration frequencies | | | | |
	Inversion segment	Interstitial segment	N	BF	LF	BBFF	LLFF
1	0	any	1	0	0	0	0
2	1	0	0	1	0	0	0
3	1	1	0	$\frac{1}{2}$	$\frac{1}{2}$	0	0
4	1	2	0	$\frac{3}{4}$	$\frac{1}{4}$	0	0
5	2	0	$\frac{1}{4}$	$\frac{1}{2}$	0	$\frac{1}{4}$	0
6	2	1	$\frac{1}{4}$	$\frac{1}{4}$	$\frac{1}{4}$	$\frac{1}{8}$	$\frac{1}{8}$
7	2	2	$\frac{1}{4}$	$\frac{3}{8}$	$\frac{1}{8}$	$\frac{1}{8}$	$\frac{1}{8}$
in 1,500 cells of *Agave stricta*			0.5150	0.3585	0.1199	0.0053	0.0013
in 318 cells of *Lilium formosanum*			0.208	0.327	0.311	0.085	0.069

Comparison with Table 2.18:	Chiasmata	0	1	2
	Observed at diakinesis	0.468	0.439	0.093
	Expected from A I observ.	0.616	0.330	0.054

categories are readily derived by sketching the different chiasma combinations in a pairing configuration and tracing the course of the chromatid from centromere to end. Chiasma frequencies higher than two are quite scare and neglected here. It is clear that for estimating the frequency of chiasmata in the inversion segment, it is not enough simply to count bridges. The first step is the estimation of the frequency of cells with one or more chiasmata in the inversion segment. In principle, this is the total minus the group without bridges (normal cells): $T - N$, but among the "normal" cells some have their origin in reciprocal chiasmata. When derived from Category 6, these may be equalled to the LLFF group, multiplied by two (i.e. $\frac{1}{4}$ vs $\frac{1}{8}$). When derived from Category 5, they equal the frequency of BBFF. In the total BBFF, however, a proportion is derived from Category 6, and this proportion equals LLFF. Therefore, together the "normals" due to reciprocal chiasmata in the inversion equal 2 LLFF + (BBFF − LLFF) or BBFF + LLFF. This has to be subtracted from the N group. The number of cells with one or two chiasmata in the inversion thus becomes $T - N$ $- (BBFF + LLFF)$. According to BRANDHAM (1969) the frequency of cells with chiasmata (1 or more) in the inversion is

$$\frac{T-N-(BBFF+LLFF)}{T}.\tag{2.4}$$

The number of cells with two chiasmata in the inversion can also be estimated. When Category 7 is neglected, they equal the sum of Categories 5 and 6. Category 6 can be assumed to be $8 \times LLFF$ and Category 5 to be $4 \times BBFF$, but here again the BBFF group derived from 6 must be subtracted; it equals LLFF. Cells with two chiasmata in the inversion, therefore, number 8 LLFF $+4(BBFF-LLFF)=4(BBFF+LLFF)$, frequency

$$\frac{4(BBFF+LLFF)}{T}.\tag{2.5}$$

Consequently the number of cells with one chiasma in the inversion equals $T-N+BBFF+LLFF-4(BBFF-LLFF)=T-N-3(BBFF+LLFF)$, frequency

$$\frac{T-N-3(BBFF+LLFF)}{T}\tag{2.6}$$

and the total number of chiasmata in T cells is $T-N+5$ $(BBFF+LLFF)$. The average number of chiasmata in the inversion is

$$\frac{T-N+5(BBFF+LLFF)}{T}.\tag{2.7}$$

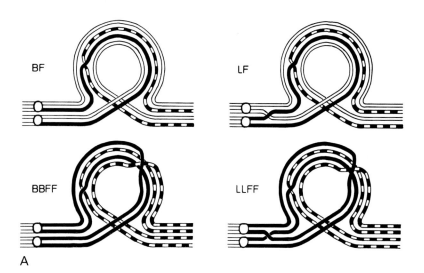

A

Fig. 2.16. (A) Some of the combinations of chiasmata in the pairing loop and in the proximal segment of a paracentric inversion heterozygote together with a reference to the anaphase configurations to be expected (bridge+fragment, *BF*; loop+fragment, *LF*; two bridges+two fragments, *BBFF*; two loops+two fragments, *LLFF*). (B) The corresponding anaphase configurations (*BF*, *LF*, *BBFF*, *LLFF*) in *Agave stricta*. (BRANDHAM, 1969)

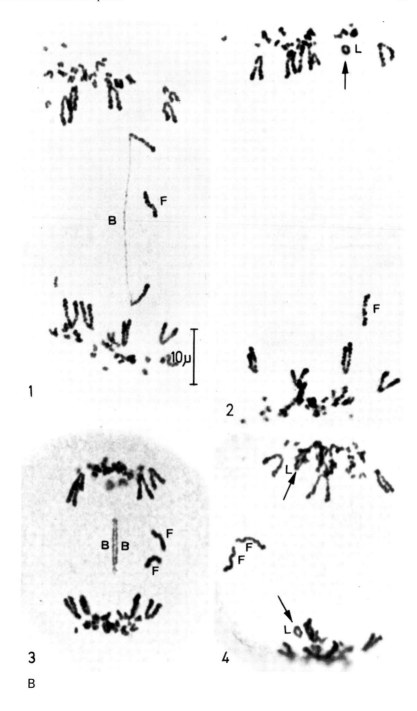

For his data (Table 2.21) from an inversion heterozygote in *Agave stricta*, BRAND-
HAM (1969) estimated that 49.2% of the cells had one or two chiasmata in
the inversion. The Categories 5 and 6 are quite small, and Category 7 may
be assumed to be negligible. It cannot be distinguished from 6, nor can any
Category with 3 or more chiasmata be distinguished. Applying Eqs. 2.5, 2.6
and 2.7 to BRANDHAM'S data gives a frequency of $\dfrac{40}{1,500} = 0.0267$ cells with two
chiasmata in the inversion, and 0.4653 with one according to Eq. (2.6). The
average number $\dfrac{1,500 - 772 + 50}{1,500} = 0.5187$.

The restrictions on estimating numbers of chiasmata in the proximal segment
are somewhat more severe: Category 4 must be negligible, as it interferes with
the separation of categories with and without chiasmata in the proximal segment.
In most materials this is not of much consequence. The calculations then amount
to adding Categories 3 and 6, but these must first be distinguished from the
others. For Category 3, double the frequency of LF is taken, but first the
contribution from 6 must be subtracted from total LF. This equals 2 LLFF,
so a value corresponding to 4 LLFF must be subtracted from LF to obtain
Category 3. Category 6 equals 8 LLFF. Thus together the number of proximal
segments with a chiasma (under the assumption that more than one do not
occur) is 2 LF − 4 LLFF + 8 LLFF = 2 LF + 4 LLFF. For finding the average
number per segment, this must be divided by the total. This is not the total
number of cells (here 1,500): since in absence of chiasmata in the inversion
no chiasmata in the proximal segment are detected, the total used is the sum
of the cells with chiasmata in the inversion, or T − N + BBFF + LLFF. The
average number of cells with a chiasma in the proximal segment therefore
is

$$\frac{2\,LF + 4\,LLFF}{T - N + BBFF + LLFF}. \tag{2.8}$$

Applying this to the data of Table 2.21, BRANDHAM (1969) found a percentage
of 49.85%, or a frequency of 0.50. In respect to genetic length the proximal
and inversion segments were apparently equal.

Comparable data on a paracentric inversion in *Lilium formosanum* were
published by BROWN and ZOHARY (1955). Processed like BRANDHAM'S agava
data they are also given in Table 2.21. There is a reasonable correspondence
with the diplotene frequencies (Table 2.20).

In *Agava stricta* no physical measurements on the inversion were possible,
but BRANDHAM estimated the genetic and physical length of the proximal and
inversion segments in *Camnula pellucida* from data presented by NUR (1968).
He reports a ratio of 10.8:1 for genetic length of proximal segment:inversion
segment, whereas the physical ratio at pachytene was 3.9:1. Apparently there
is chiasma localization near the centromere. An alternative explanation is that
reduced effectiveness of pairing has led to lowered chiasma frequencies in the
inversion segment.

In principle, it is possible to detect and even estimate interference between the proximal and inversion segments. The cause of interference may be purely mechanical. One can visualize a reduction of pairing and consequently of chiasma formation in the inversion segment when the proximal segment has paired completely and *vice versa*. Since there are five types of configuration (N, BF, LF, BBFF and LLFF) with a sum equal to 1, giving four degrees of freedom, there is one degree of freedom left for interaction when three parameters have been estimated (frequency of 1 chiasma in the proximal segment, frequency of 1 chiasma in the inversion segment, frequency of 2 chiasmata in the inversion segment). In the equations given above, this fourth degree of freedom has not been included, since in all cases BBFF and LLFF have been used together, as their sum. The required additional information for interference, therefore, cannot be derived from the equations given so far, nor from the distributions derived from them. In fact it can be seen that when all expected values are calculated for the individual contributions of the different Categories 1–6, excluding 4 and 7, they add to the observed frequencies of the five configurations: all information is used (Table 2.21). Only by separating BBFF and LLFF can the interaction, i.e. interference, be estimated. It can be intuitively understood that without interference between the proximal and inversion segments the ratio BF/LF must equal BBFF/LLFF. It can also be derived more exactly. For convenience of notation renaming the categories $2 \rightarrow a$; $3 \rightarrow b$; $5 \rightarrow c$; $6 \rightarrow d$, one can state that $\dfrac{a + \frac{1}{2}b + \frac{1}{2}c + \frac{1}{4}d}{\frac{1}{2}b + \frac{1}{4}d} = \dfrac{\frac{1}{4}c + \frac{1}{8}d}{\frac{1}{8}d}$. This is readily shown to be true when $\dfrac{a}{b} = \dfrac{c}{d}$, or, when a chiasma in the proximal segment is formed as frequently when there are two chiasmata in the inversion, as when there is only one. It may be noted that the category "0 chiasmata in the inversion, chiasmata in the proximal segment" is not available for the comparison, as without chiasmata in the inversion the presence of chiasmata in the proximal segment is not detected. It thus appears that interference between proximal and inversion segments is estimated by comparing the ratio BF/LF with BBFF/LLFF. The first is approximately 2.9, the latter 4. Since only 2 LLFF were observed, the difference is not at all significant: with 3 LLFF the ratio BBFF/LLFF would have been 2.7. All that can be said is that if there is any interference at all, it cannot be very strong. Apparently the detection of interference requires the inversion to be large enough to have frequently two chiasmata. This is one consequence of the fact that the comparison between 1 chiasma and 0 chiasma in the inversion cannot be used in reference to chiasmata in the proximal segment.

When diplotene-diakinesis and anaphase data are available of the same material, they can be compared. For the paracentric inversion in *Lilium formosanum*, reported by BROWN and ZOHARY (1955), the correspondence is somewhat disappointing (Table 2.21).

An interesting combined analysis of meiotic observations and marker segregations in paracentric inversion heterozygotes in barley was reported by EKBERG (1974) who developed equations relating the two types of data.

2.2.3.2 Pericentric Inversions

With pericentric inversions, the centromere is in the inverted segment. Pairing between the normal and the inversion chromosome in a heterozygote again leads to the formation of a pairing loop. A chiasma in this loop is a reversed chiasma, as with a paracentric inversion, and can similarly be scored at diplotene or diakinesis. Again, the symmetry of the configuration is determined by the location of the chiasma in the pairing loop (Fig. 2.14, cf. ZOHARY, 1955). The analysis is entirely comparable with that in paracentric inversions.

A chiasma in the pairing loop does not lead to an anaphase I bridge in pericentric inversions, but to deficiency-duplication chromatids which can differ in length to such an extent that they become recognizable at anaphase I, or at metaphase II and anaphase II (Fig. 2.17). The larger the difference between the normal chromatids and the deficiency and duplication chromatids, the more easily are the latter recognized and, therefore, the easier it is to score crossing-over in the inversion. This difference is independent of the location of the chiasma and of the location of the centromere inside the pairing loop. It depends only on the difference between the two segments outside the inversion, which it equals. The analysis of crossing-over inside the inversion is not simpler than that in paracentric inversions, but chiasmata outside the inversion do not interfere with the analysis. Usually the comparison between normal chromatids and deficiency-duplication chromatids is easiest at first anaphase, where they occur

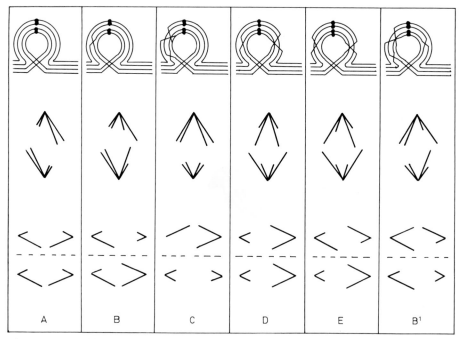

Fig. 2.17A–E. Pairing, and anaphase I and anaphase II configurations with different combinations of chiasmata in the pairing loop of a pericentric inversion heterozygote

side by side. In a number of situations, however, notably those of Fig. 2.17 D and E, it is difficult to see which chromatids are attached to the same half-centromere.

Then anaphase II is a more reliable source of information, because the chromatids have already separated. A disadvantage, however, is that fewer different types can be distinguished than in A I and consequently, that the number of degrees of freedom available for estimating different parameters is reduced.

In plants, where the two daughter cells remain associated, all anaphase II chromatids of the same sporocyte can be compared and scored. In animals the two daughter cells usually separate before metaphase II and the two products of the same AI event cannot be analyzed in combination. As seen in Fig. 2.17, this is not of much consequence: the two secondary spermatocytes of type B are unique and can always be recognized. If there is a difference in frequency, this should merely be due to sampling error. The same is true for type C. In type E, however, one secondary spermatocyte resembles A and the other resembles D. As will be shown later, there is a constant ratio of 2 between types E and D, and therefore of all observed secondary spermatocytes of type D, half actually have originated as type E, and a similar number must be subtracted from type A to find the number of type A proper. Except for an increase in sampling error, separation of the two spermatocytes does not lead to complications in the analysis of crossing-over in the inversion.

It was again BRANDHAM who, in 1970, reported on a quantitative analysis of crossing-over in a pericentric inversion, based on anaphase II data from Gasteria, collected by GILES (1943). Like the Aloineae in general, Gasteria has eight long sub-acrocentric chromosomes with many chiasmata in each, and six small ones. The pericentric inversion studied by GILES had occurred in a long chromosome and although of considerable size, was not long enough to cause striking alteration in the shape of the chromosome. The original interpretation, that the newly formed metacentric chromosomes at A II were isochromosomes resulting from centromere misdivision had already been abandonned by GILES in 1944 in favor of an interpretation based on chiasma formation in a pericentric inversion heterozygote. Although BRANDHAM'S aim was primarily to work out a more quantitative argumentation for the same conclusion, his approach can be used to estimate crossing-over in the inversion, but it must be carried somewhat further than BRANDHAM needed for his purpose.

Only anaphase II classifications were given by GILES. Fig. 2.17 shows different pachytene, anaphase I and anaphase II configurations with 0, 1, 2 and 3 chiasmata in the loop. With one chiasma, irrespective of position and chromatids involved, anaphase category B will result. With two chiasmata, it makes a difference which chromatids are involved (complementary, reciprocal, disparate 1, disparate 2), and in addition their location in respect to the centromere is of importance (Fig. 2.17 C, D and E). In D two complementary chiasmata, each on one side of the centromere are shown, in E two disparate chiasmata. Disparate 1 and disparate 2 give identical results. Reciprocal chiasmata give a distinct anaphase I type (two groups of the type at the top in E) but in anaphase II are indistinguishable

Table 2.22. The theoretical frequencies of the five A I (or A II) configuration types (A, B, C, D, E) in the pericentric inversion heterozygote of Fig. 2.17 for different numbers and combinations of chiasmata (m, n, o, p, r) in the inversion. (BRANDHAM, slightly extended, 1970). The observations on Gasteria are by GILES, 1943

Chiasmata			Expected frequency of configuration types of Fig. 2.17				
Number	Position		A	B	C	D	E
0	—	(m)	1	—	—	—	—
1	anywhere in inversion	(n)	—	1	—	—	—
2	both sides of centromere	(o)	$\frac{1}{4}$	—	—	$\frac{1}{4}$	$\frac{1}{2}$
2	same side of centromere	(p)	$\frac{1}{4}$	$\frac{1}{2}$	$\frac{1}{4}$	—	—
3	2 on one side, 1 on other side	(r)	$\frac{1}{8}$	$\frac{1}{2}$	—	$\frac{1}{8}$	$\frac{1}{4}$
In 184 cells of *Gasteria maculata*							
Number			35	98	13	10	28
Frequency			0.191	0.532	0.071	0.054	0.152

from A. A review of all 5 possible anaphase II categories and their derivation is given in Table 2.22 (an expansion of BRANDHAM'S 1970 table) together with the frequencies reported by GILES (1943). The sum is 1 and D and E have a fixed ratio of 2, and consequently there are only three degrees of freedom. Our first interest is estimating the frequencies of the different combinations of chiasmata indicated by m, n, o, p, r. They can be expressed in terms of A, B, C, D, E on the basis of Table 2.22:

$$A = m+\tfrac{1}{4}o+\tfrac{1}{4}p+\tfrac{1}{8}r = 0.191$$
$$B = n+\tfrac{1}{2}p+\tfrac{1}{2}r \qquad = 0.532$$
$$C = \tfrac{1}{4}p \qquad\qquad\quad = 0.071$$
$$D = \tfrac{1}{4}o+\tfrac{1}{8}r \qquad\quad = 0.054$$
$$E = \tfrac{1}{2}o+\tfrac{1}{4}r \qquad\quad = 0.152.$$

The ratio E/D does not deviate significantly from 2:1; observed 28 E and 10 D; expected 26 E and 13 D. When leaving out the possibility of 3 chiasmata in the inversion, there are enough degrees of freedom, and we find that $m=0.051$; $n=0.390$; $o=0.276$ and $p=0.284$. Together these represent three independent parameters, the sum being 1. This will quite frequently be sufficient. In this particular example, however, there is every reason to suspect that in rather a large number of cells two chiasmata will have occurred on one side of the centromere, together with one at the other side (combination r). This is based on the high values found for *both o and p*.

The anaphase II information is insufficient to estimate a fourth parameter, so recourse must be made to outside information. In this case, the distribution of the chiasmata, although not accurately predictable for any specific situation, may be used. First, the frequency of the inversion being "bound" by at least one chiasma is estimated. From $A = m+C+D$, m is found to be 0.066. This is the frequency "not bound"; $(1-m)=0.934$ is the frequency of the inversion being "bound". From this (using the proper mapping function) the average frequency of crossing-over is calculated. Since this value is actually too high for any mapping function, none of the functions available is entirely satisfactory, and we therefore use KOSAMBI'S, which is the simplest to work with. It yields a crossing-over frequency of 0.85 or an average chiasma frequency of 1.7. It is not well possible to use this to calculate the frequency of three chiasmata occurring in the inversion. If the distribution of chiasmata had been random, the Poisson series could have been used, with 0.34 as the frequency of three chiasmata. Now a process of iteration (or, rather, trial and error) is most conveniently applied. If we take 0.20 for r, and use this in the five equations, the values for n, o and p are solved as $n=0.290$; $o=0.160$; $p=0.284$. Now the average chiasma frequency in the inversion is $0.290 \times 1 + 0.160 \times 2 + 0.284 \times 2 + 0.2 \times 3 = 1.78$. This is higher than the value of 1.7 given above. With $r=0.15$; $n=0.315$; $o=0.185$ and $p=0.284$, the average chiasma frequency equals 1.70, which is as expected (KOSAMBI function).

In view of the relatively high frequency of cells with 2 chiasmata at one side of the centromere and the much lower frequency with one on either side, it is probable that the centromere is located close to one side of the loop. Then practically all cases with two chiasmata on one side of the centromere will have these chiasmata in the long segment, the short segment practically never having two. Combination n consists of two types: that with one chiasma in the long segment and none in the short one (n^a) and that with no chiasma in the long segment but one in the short (n^b). It may be assumed that n^a is much larger than n^b. The ratio between o and n^a must equal that between r and p when there is no interference across the centromere. The derivation of r, however, is based upon KOSAMBI'S mapping function in which a considerable degree of interference is assumed. Indeed, from the values of o and n it is clear that the ratio between o and n^a can never equal that between r and p: apparently, when there are two chiasmata in the long segment, a chiasma in the short segment is much less frequent than when there is only one chiasma in the long segment. Since the long segment is rather large, the effect of a single chiasma compared to no chiasma in the long segment may be expected to be less pronounced, so that it is not unreasonable to assume n^b to equal o, or to be slightly larger. In addition, also for interference reasons, n^b should be larger than o. This is irreconcilable. Rather arbitrarily we can guess $n^a=0.160$ and $n^b=0.155$.

The following can now be calculated as averages:

Chiasmata in the long
segment: $0.160 + 0.185 + 0.568 + 0.300 = 1.213$; map length 0.607

Chiasmata in the short
segment: $0.155 + 0.185 + 0.150$ $= 0.490$; map length 0.250

                                     Total   1.703   0.857

Chiasma distribution:

Long segment		Short segment
Chiasmata	Frequency	Frequency
0	0.221	0.510
1	0.345	0.490
2	0.434	0.000
3	0.000	0.000
	1.000	1.000

The analysis has become progressively speculative, and is somewhat unsatisfactory in its final conclusions mainly because of: (a) small sample; (b) the use of KOSAMBI'S function for a map length to which it is not suited; (c) guessing the division of n into n^a and n^b.

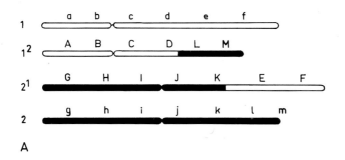

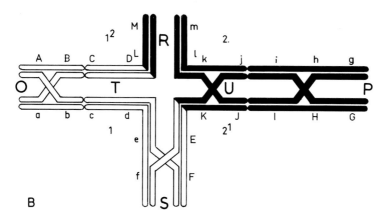

Fig. 2.18A and B. The chromosomes and the pairing configuration (with four chiasmata in arbitrary positions) of a reciprocal translocation. The same segments O, P, R, S, T, U as in a tetrasomic pairing configuration with one point of partner exchange (Figs. 2.9; 2.11) can be distinguished but here the point of partner exchange is fixed and in principle coincident with the break point

2.2.4 Reciprocal Translocations

2.2.4.1 Direct Estimates of Chiasma Frequencies

Reciprocal translocations are the most common and also the most easily recognized and analyzed type of translocation. They arise as a result of an interchange of segments between two non-homologous chromosomes and are therefore often called *interchanges*.

In the heterozygote, when all homologous segments have paired at pachytene, the cross-shaped configuration of Fig. 2.18 is formed. When the chromosomes are not telocentric there are six segments: the two uninvolved arms O and P; the two translocated segments R and S, and the two interstitial segments T and U. With strongly acrocentric or telocentric chromosomes, there are only the four segments R, S, T, U. Depending on the formation or non-formation of chiasmata in these segments, different configurations will appear at diakinesis or metaphase I, after opening up of the pairing cross (Figs. 2.18; 2.19; 2.20). The

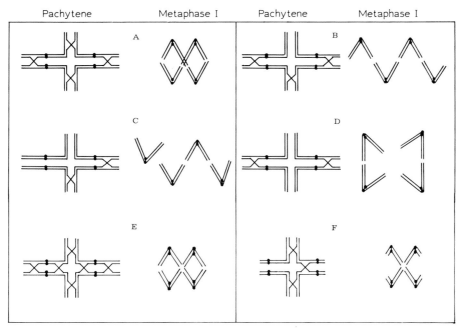

Fig. 2.19A–F. Examples of pairing and M I configurations (alternate orientation) with different combinations of chiasmata in a reciprocal translocation. (A–E) Metacentric chromosomes. (F) Acrocentric chromosomes. All chiasmata are assumed to have terminalized completely, even across the translocation break point

special characteristics of such configurations help in identifying the segments in which chiasmata have formed, but often there is so much symmetry in the complex that distinction between O and P, between R and S and between T and U

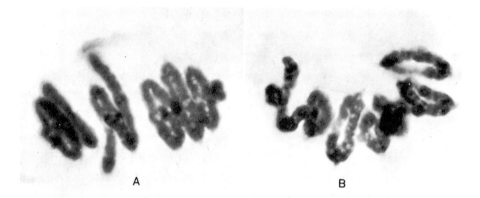

Fig. 2.20A and B. Two M I configurations of a reciprocal translocation heterozygote
in rye. (A) Chain quadrivalent (cf. Fig. 2.19B). (B) "Frying-pan" quadrivalent as is formed
with the combination of chiasmata of Fig. 2.18B. (SYBENGA, 1972a)

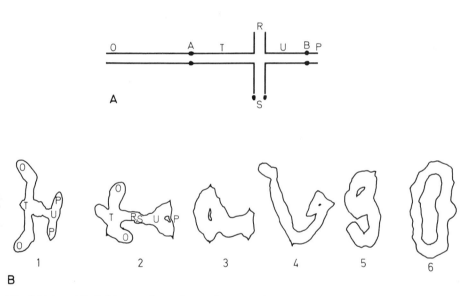

Fig. 2.21A and B. Pairing configuration and M I configurations of a reciprocal translocation
heterozygote in *Delphinium ajacis*. As a result of the difference in size between the (A)
and (B) sets of chromosomes the individual segments of the quadrivalent can be recognized
at M I. In all examples shown at least one of the two translocated segments has a (completely
terminalized) chiasma, which is a prerequisite for keeping the quadrivalent together (cf.
Fig. 2.18). In (B1), both interstitial segments T and U have a chiasma, in addition to
one of the translocated segments (R or S), but none of the unaffected arms (O, P). In
(B2), the same holds in respect to the interstitial and translocated segments, but there
is also a chiasma in the small arm P. In (B3), the large arm, O, has a chiasma in addition
to those in (B2). In (B4), there are chiasmata in the small arm P, in one translocated
segment and in interstitial segment U. In (B5) the same situation holds, but the interstitial
segment with the chiasma is T. In (B6) neither interstitial segment has a chiasma, but
all four remaining segments have one. (Drawn in outline from photomicrographs in JAIN
and BASAK, 1963)

Table 2.23 (A). Number and frequency of pollen mother cells with and without chiasmata in the two interstitial segments a and b at MI in 300 cells of a translocation heterozygote of *Delphinium ajacis*, and the A I segregation in 183 cells. Cf. Fig. 2.19 (JAIN and BASAK, 1963). The interstitial segments can be distinguished at M I, as well as the two chromosome pairs at A I. A and B are the normal, A_1 and B_1 the translocation chromatids. An interstitial chiasma leads to the combination of an A with an A_1 chromatid

	Metaphase I				Anaphase I			
	Segment a		Segment b		A pair		B pair	
	0 chiasma	1 chiasma	0 chiasma	1 chiasma	$AA-A_1A_1$ 0 chiasma (reductional)	AA_1-AA_1 1 chiasma (equational)	$BB-B_1B_1$ 0 chiasma (reductional)	BB_1-BB_1 1 chiasma (equational)
Cells	166	137	130	170	98	85	84	99
Frequency	0.553	0.447	0.433	0.567	0.536	0.467	0.459	0.541

Table 2.23 (B). Chiasma frequency at M I (1974 cells) and segregation at A I (478 cells) in a translocation heterozygote of *Scilla scilloides* (NODA, 1961)

	Metaphase I				Anaphase I			
Year	Chiasmata in interstit. segm.			Cells	0 chiasma both (reduct. both)	1 chiasma either (1 reduct. 1 equat.)	1 chiasma both (equat. both)	Cells
	0 in both	1 in either	1 in both					
1958	80 0.100	276 0.345	444 0.555	800	6 0.056	38 0.351	67 0.593	108
1959	144 0.123	404 0.344	626 0.533	1174	40 0.108	148 0.400	182 0.492	370
total	224 0.113	680 0.344	1070 0.543	1974	46 0.096	186 0.389	246 0.515	478
Chiasmata per segment	0.714 M I				0.709 A I			

Table 2.23C. Frequencies of chiasmata in the unequal arm of a heteromorphic bivalent in *Lilium callosum* (KAYANO, 1959), and expected and observed frequencies of equational and reductional AI segregation. (A) no interference; (B) complete chromatid interference

Chiasmata				AI segregation (70 cells)					
number per arm	observed		Poisson freq.	equational exp.		obs.	reductional exp.		obs.
	number	freq.		A	B		A	B	
0	0	0	0.1405	—	—	—	—	—	—
1	50	0.235	0.2757	0.235	0.235	—	—	—	—
2	121	0.568	0.2705	0.284	—	—	0.284	0.568	—
3	42	0.197	0.1769	0.148	0.197	—	0.049	—	—
4	0	0	0.0867	—	—	—	—	—	—
5	0	0	0.0340	—	—	—	—	—	—
total	213	1.000	(0.9844)	0.667	0.432	0.486	0.333	0.568	0.514

is not possible. Only when at least one of the three sets of two symmetrical segments is marked (for instance by being conspicuously different in size), is individual recognition of the six segments possible.

2.2.4.1.1 Metacentric Chromosomes

An example of a reciprocal translocation between two metacentric chromosomes of *Delphinium ajacis* is given by JAIN and BASAK (1963).

Segment O is much larger than P and the two are easily distinguished (Fig. 2.21). As a result, T, which is adjacent to P, even when similar in length to U, can still be recognized and so can R and S. For the interpretation of the configuration it is useful to draw the attention to the fact that one of the two exchanged segments is quite small and usually has no chiasmata. The other is larger and usually has one. This chiasma tends to become terminalized. In this particular example, the original purpose was not to analyze chiasma frequencies in all six segments, but to concentrate on the interstitial segments, which incidentally were of rather different size, and to compare chiasma frequencies in the interstitial segments with the frequency of observed chromatid exchange at anaphase I. Table 2.23A shows the results. They are relatively simple: there was never more than one chiasma in each segment.

The data permit an analysis of interference between the two interstitial segments. The frequency at metaphase of chiasmata was 0.447 in one segment, in the other segment it was 0.567; the expected frequency of a chiasma in both simultaneously, therefore, is 0.253. Chiasmata in both interstitial segments simultaneously were observed in 0.287 of the cells. Expressed as numbers per cell a comparison with the numbers observed can be made:

	expected	observed	
chiasma in both	75.9	86	
chiasma in neither	64.7	86	χ_2^2 14.5
chiasma in either	159.4	128	
	300	300	

The difference is significant; there is interference between the two interstitial segments. It is *negative*, which is important to note, since this will be encountered again, and tentatively explained later in this discussion. Very similar results were presented by NODA (1960) on a *Scilla scilloides* interchange heterozygote. Here the two non-translocation arms and the two interstitial segments were so similar that no distinction was possible: chiasmata in the two interstitial segments had to be pooled. The interchanged segments, however, differed considerably, and therefore crossing-over in the interstitial segments lead to clearly different chromatids in the anaphase I chromosomes. The observations, covering two consecutive years, are shown in Table 2.23 B. There was, again, no case of more than one chiasma in each segment studied. Since, however, the two segments cannot be distinguished, a direct analysis of interference is not possible. All that could be scored was: both interstitial segments are bound; only one is bound; neither is bound. It is a similar situation as in the case of normal bivalents with two arms that cannot be distinguished: there are three categories with a sum equal to 1, and thus only two degrees of freedom, allowing estimation of only two parameters. There are three: the two chiasma frequencies for the two interstitial segments and their interaction (interference). In the case of the normal bivalent it was often reasonable to assume absence of interference, and estimates for the two arms could be derived. In the case of the interstitial segments of a translocation, however, since our interpretation of the data of JAIN and BASAK (1963) suggests strong negative interference, we must be cautions with any assumptions about the absence of interference. It is equally impossible to make *a priori* statements about the chiasma frequencies themselves. There is no reason for instance, to assume that they are equal. There remains the possibility of *detecting* negative interference, and its level above the threshold of detection. This has been considered before when discussing the method of solving the quadratic equations for the frequencies of being bound for the two arms of normal bivalents. It is based on the necessity of the discriminant to be positive. In the present example, the frequency of both interstitial segments having a chiasma, $t \cdot u = 0.542$. One segment bound, the other not, has a frequency of $t + u - 2tu = 0.344$. This leads to the roots $x_{t,u} = \dfrac{1.428 \pm \sqrt{2.039 - 2.168}}{2}$. The discriminant is negative, which suggests negative interference. A statistical analysis is not well possible, but the statistical significance of this statement can be approached in a special way: with an average frequency of being bound of 0.714 per segment, the frequency of simultaneous occurrence of chiasmata in both segments is expected to be $0.714^2 = 0.5098$ when the segments are equal; simultaneous absence $(1 - 0.714)^2 = 0.0818$; a chiasma in one segment and none in the other $2 \times 0.714 \, (1-0.714)=0.408$. In a total of 1,974 segment pairs this corresponds with the following distribution.

	Expected	Observed	
1 in both	1,006.3	1,070	
1 in either	806.2	680	$\chi_2^2=22.6$
1 in neither	161.5	224	
	1,974	1,974	

The difference is highly significant, indicating that even when both segments are equal, negative interference is beyond doubt. When the two segments are unequal, which they probably are, the difference is even more pronounced. After estimating interference, there still is a degree of freedom left, which may be used to estimate the average frequency of being bound, or the sum, which is $t+u=0.344+2\times0.542=1.428$. It may be noted in passing that in this particular case, where the maximum number of chiasmata is 1, the frequency of being bound coincides with the chiasma frequency. Conversion into map length with the use of mapping functions is now unnecessary: map length equals $\frac{1}{2}$ chiasma frequency $\times 100 = 35.7$ units.

Several more similar reports have been published which can be analyzed along these lines, but the examples given may suffice. In the cases in which in addition to information on the interstitial segments also chiasma frequencies for the other segments (O, P, R, S) are given, it is usually not clear if actually chiasmata have been counted or if merely *association by chiasmata* is given as the chiasma frequency. Usually it is suspected that the latter is the case, and therefore these cases will be considered in Section 2.2.4.2.

2.2.4.1.2 Acrocentric Chromosomes

Compared to those in metacentric chromosomes, heterozygous reciprocal translocations in acrocentric chromosomes have a decreased number of possible diakinesis and metaphase I configurations (Fig. 2.22). There are four instead of six segments in which chiasmata can be formed: two interchanged and two interstitial segments. When at least one of the constituent chromosomes is recognizable and included in a multivalent, and when the centromeres can be distinguished, the different segments in the multivalent can be recognized and the number of chiasmata in each of them counted. When the recognizable "marker chromosome" is univalent, the order of the chromosomes in the remaining trivalent cannot be determined. When the marker is in a bivalent the other two chromosomes of the translocation complex cannot usually be distinguished from any other pair of the genome.

Recognizable chromosomes are not always available and, moreover, multivalents are not always formed: they require simultaneous chiasma formation in translocated and interstitial segments. The latter is realised only when both segments are relatively long, and when interference between them is not strong.

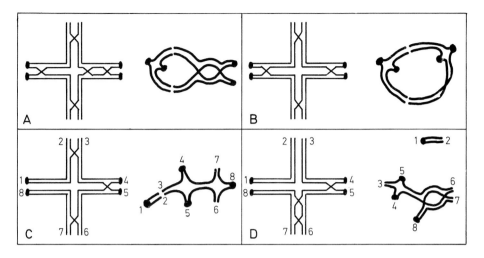

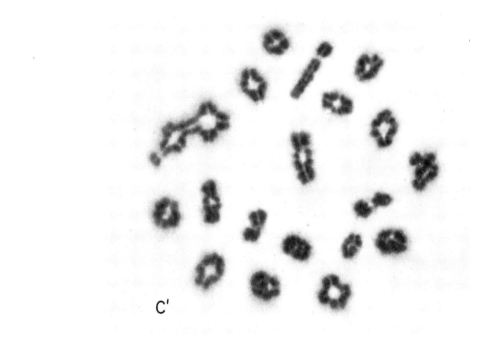

Fig. 2.22 A–D. Pairing and diakinesis configurations of a reciprocal translocation heterozygote involving acrocentric chromosomes. (A) and (B) are ring quadrivalents, (C) and (D) are chains. (C1) shows the diakinesis appearance in the mouse of diagram (C). (Photomicrograph courtesy C. E. FORD). In Fig. 2.23 (D2) the trivalent and univalent of (D) are shown at M I in *Acrida lata* (courtesy M. SANNOMIYA). Note the superficial similarity of this trivalent with that of Fig. 2.12 (B1). As appears from a comparison of Figs 2.12 (A5) and 2.22 (D), the positions of the centromeres are different in the two trivalents

Interference between interstitial and translocated segments has repeatedly been shown to be significant in various material. Without proper recognition of the segments direct estimates of chiasma frequencies are not well possible and indirect estimates have to be obtained. They are treated in Section 2.2.4.2. Often, in the most frequently used materials with acrocentric chromosomes (for transloca- tions mainly the house mouse, *Mus musculus* (cf. FORD, 1969; SEARLE et al., 1971; DE BOER, 1975) and a number of locusts and other insects) chiasma terminalization at diakinesis is limited and it is probable that the number of chiasmata observed equals the number formed. This is exemplified by SANNO- MIYA'S (1968) observations on *Acrida lata* where in a heteromorphic bivalent 1.80 chiasmata were reported at diplotene and 1.77 at MI. Therefore, even in the case of terminal chiasmata, where on the basis of the observation as such, doubt about the number of chiasmata might arise, only a single chiasma may be assumed to have been formed. Moreover, in such material, for short segments the number of chiasmata tends to be limited to one per segment. This implies that, unless the opposite is actually observed, the frequency of being bound, for the translocation and for the interstitial segments for practical purposes coincides with the frequency of chiasmata, whether this frequency has been determined by direct observation or, indirectly, from the relative frequencies of different configurations (Section 2.2.4.2).

It should be noted that for longer segments in most material with effectively terminalizing chiasmata this correspondence between the number of chiasmata formed and observed at diakinesis/metaphase is not nearly as good. There are two reasons: one is fusion of chiasmata at the chromosome end, the other is actual loss of chiasmata from the end. As stated earlier, there are numerous examples of a decrease in chiasma frequency even between diakinesis and meta- phase, and this seems to be the rule rather than the exception with terminalizing chiasmata. Nevertheless, in the examples given here the correspondence is good.

Fig. 2.22 shows a few of the several possible diakinesis configurations involving acrocentric chromosomes (cf. SEARLE et al., 1971; DE BOER, 1975). The probable interpretation in terms of presence and absence of chiasmata is also shown in a diagrammatic pairing configuration. The frequencies of the different diakinesis configurations (rings, chains, etc.) reported for the mouse are not satisfactory for direct estimates of chiasma frequencies, since the chromosome segments involved have not been specified. The data presented by SANNOMIYA (1968) on an interchange heterozygote in *Acrida lata* are more specific (Fig. 2.23).

Fig. 2.23A–D. Reciprocal translocation between two acrocentric chromosomes of *Acrida lata*. (A) The chromosomes. (B) The pairing configuration with one chiasma in the 9-9^3 interstitial segment and two in the 3-9^3 translocated segment. This combination leads to the trivalent with univalent of (D2) (cf. Fig. 2.22D). (C) Anaphase I segregation with equal chromatids in chromosomes 3 and 3^9, and unequal chromatids in chromosomes 9 and 9^3, resulting from one chiasma in the 9-9^3 interstitial segment (cf. D3, arrows). (D) *1*: M I in normal *Acrida lata* ♂; *2*: M I in translocation heterozygote, with trivalent and univalent; *3*: A I in translocation heterozygote, note unequal chromatids due to a chiasma in the interstitial segment (arrows); *4*: M II in translocation heterozygote, note unequal chromatids (arrow) as in A I; *5*: A II separating the unequal chromatids (arrows). SANNOMIYA (1968)

The interchange involved chromosomes 3 and 9. In one interstitial segment (T) no chiasma was ever observed in 217 cells. Segment S never failed to have a chiasma. In a personal communication the average number of chiasmata in S was given as 2.392. Since this is one of the few examples of a complete analysis, the data are given here *in extenso* (Table 2.24). In segment U a single chiasma was observed (at MI) in 0.981 of the cases but never two, and in segment R a frequency of 0.074 chiasmata was reported, again never two at the same time. This permits a simple analysis of interference between R and U:

	Frequency	Number expected	Number observed
Chiasmata in both R and U	$0.981 \times 0.074 = 0.0726$	15.75	14
Neither in R nor in U	$0.019 \times 0.026 = 0.0176$	3.82	2
One in U and none in R	$0.981 \times 0.926 = 0.9084$	197.2	199
One in R and none in U	$0.019 \times 0.074 = 0.0014$	0.30	2
			217

Thus in a total of 217 configurations 14 chain quadrivalents were observed (with R, S and U bound but T free); twice a bivalent with two univalents (only S bound); 199 trivalents with univalent (S and U bound but T and R free); and twice a pair of bivalents (R and S bound but T and U free). The correspondence is surprisingly good. The probable reason why interference could not be demonstrated in this particular case will be briefly discussed later (Section

Table 2.24. The number of cells with 0, 1, 2 or 3 chiasmata in the four segments of a translocation heterozygote (acrocentric chromosomes) of *Acrida lata* (SANNOMIYA, 1968). Cf. Fig. 2.23

Configurations	Cells	Chiasmata in segment:			
		R	S	T	U
Chain quadrivalent $(3-9^3-9-3^9)$	14	1 chiasma: 14	1 chiasma: 1 2 chiasm.: 9 3 chiasm.: 4	0	1 chiasma: 14
Triv.+univ. $(3-9^3-9)+(3^9)$	199	0	1 chiasma: 1 2 chiasm.: 115 3 chiasm.: 83	0	1 chiasma: 199
2 biv. $(3-9^3)+(3^9-9)$	2	1 chiasma: 2	1 chiasma: 1 2 chiasm.: 1	0	0
1 biv.+2 univ. $(3-9^3)+(9)+(3^9)$	2	0	2 chiasm.: 1 3 chiasm.: 1	0	0
Total	217	16	519	0	213
Average chiasmata/segm.		0.074	2.392	0.000	0.981

2.2.4.2.3). Interference between S and U cannot be estimated since the number of cells in which U did not have its chiasma amounted to four only. Interference between S and R offers better opportunities for analysis. Segment R either had one chiasma or none. Table 2.24 shows that in the 16 cells in which R did have a chiasma, S had a total of 34. In the 201 cells in which R did not have a chiasma, S had 485 chiasmata. A test using a 2×2 table gave a χ^2 of 0.174, which indicates independence.

2.2.4.2 Crossing-over Estimated from Metaphase Association

2.2.4.2.1 Metacentric Chromosomes

In most translocation heterozygotes recognition of a specific "marker" chromosome permitting the identification of the different associated segments at meiosis is not possible. Usually, morphological differentiation between the chromosomes is insufficient, even when at mitosis the four chromosomes of the complex are distinct. Or recognition is too uncertain in a significant proportion of the meiotic configurations to permit a meaningful analysis of crossing-over in specified segments. Then, even when the chiasmata can be observed individually in all six segments, there is no means of assigning the observed number of chiasmata to a particular segment. In a number of configurations (notably in those with interstitial chiasmata) the two non-translocation arms can be distinguished from the translocated and interstitial segments on the basis of the general characteristics of the configuration. However, such configurations in any case are only a fraction of the total. There are methods of analysis, comparable to those used for normal metacentric bivalents with morphologically indistinguishable arms (Section 2.1.2.2), which make it possible to estimate the frequency of being bound for the six different segments, even when specific segments cannot be recognized. Because of symmetry in the configurations it is usually not possible to distinguishe specifically between two equivalent segments, although a separate value can be extracted for each of the two. Since these methods are based on the relative frequencies of characteristic diakinesis and metaphase I configurations, which result from the association of specific segments, no use can be made of chiasma frequencies in such segments, simply because these are not recognized as such. This is not too serious, as it is usually impossible to determine these chiasma frequencies sufficiently accurately in complex configurations.

In order to develop a system for estimating frequencies of being bound from observed configuration frequencies, it is useful first to establish which configurations result from specific combinations of bound and not-bound segments. This will first be done for metacentric (and submetacentric) chromosomes, which are the most "complete" chromosomes, and then for acrocentric chromosomes which can be treated as a simplified case. As seen in Fig. 2.18 a heterozygote for a reciprocal translocation between metacentric chromosomes has six segments, O, P, R, S, T, U. In these segments, chiasmata may either be present of absent.

With these two possibilities for each segment, there are 2^6 or 64 combinations for the six together. Each combination corresponds with one diakinesis or metaphase I configuration, but since the different chromosomes in the configurations usually cannot be recognized, a number of these configurations have the same general shape. A total of sixteen different characteristic shapes can be distinguished (Fig. 2.24; cf. Fig. 2.19). Of these, 11 are multivalents and five are combinations of bivalents and univalents. The latter five are usually so similar in shape to the bivalents and univalents formed by the remainder of the genome that in general their frequency cannot reliably be determined. If necessary, an approximate estimate can be made when the frequencies of ring and open bivalents and univalent pairs formed by the non-translocation chromosomes are known. For a first analysis only the multivalents, eleven types, are available. The frequency of each of these configurations can in principle be represented by the product of the probabilities of each segment having a chiasma or not, depending on what applies for the particular configuration, cf. Figs. 2.18; 2.19. For instance,

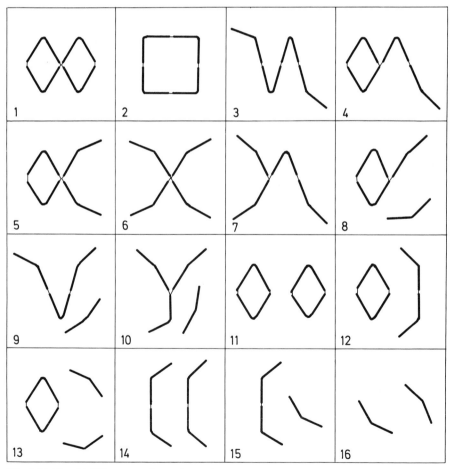

Fig. 2.24. The 16 different configurations of a heterozygous reciprocal translocation between metacentric chromosomes (cf. Table 2.25 and Figs. 2.18, 2.19, 2.20, 2.21)

the frequency of a ring of four chromosomes can be given as $o \cdot p \cdot r \cdot s \cdot (1-t) \cdot (1-u)$ where o, p, r, s, t and u are the frequencies of being bound for the segments O, P, R, S, T and U respectively. This, then, is the configuration in which O, P, R and S are bound, but T and U not. It is the only way in which this configuration can arise. The chain of four chromosomes, however, has four different origins; failure of chiasmata either in O or in P or in R or in S combined with failure of chiasmata in T and U all lead to a chain of four. The frequency of this configuration, therefore, is:

$$(1-o) \cdot p \cdot r \cdot s \cdot (1-t) \cdot (1-u) + o \cdot (1-p) \cdot r \cdot s \cdot (1-t) \cdot (1-u)$$
$$+ o \cdot p \cdot (1-r) \cdot s \cdot (1-t) \cdot (1-u) + o \cdot p \cdot r \cdot (1-s) \cdot (1-t) \cdot (1-u).$$

In this way the frequency of each configuration type can be expressed in terms of o, p, r, s, t and u (Table 2.25 and Fig. 2.24).

Most conveniently, when working out the functions expressing the frequencies of certain configurations, for translocations or otherwise, all possible combinations (64 in this case) are written out separately and the resulting diakinesis or metaphase I configurations constructed. Then all configurations of the same type are combined and the corresponding frequency functions added.

Table 2.25. The expected frequencies of the configurations of Fig. 2.24 (heterozygote of a reciprocal translocation between two metacentric chromosomes), in terms of the probabilities o, p, r, s, t, u of the six segments O, P, R, S, T, U being bound by at least one chiasma

1. $o \cdot p \cdot (1-r) \cdot s \cdot t \cdot u \cdot$
   $+ o \cdot p \cdot r \cdot (1-s) \cdot t \cdot u \cdot$
   $+ o \cdot p \cdot r \cdot s \cdot (1-t) \cdot u \cdot$
   $+ o \cdot p \cdot r \cdot s \cdot t \cdot (1-u)$
   $+ o \cdot p \cdot r \cdot s \cdot t \cdot u$

2. $o \cdot p \cdot r \cdot s \cdot (1-t) \cdot (1-u)$

3. $o \cdot p \cdot r \cdot (1-s) \cdot (1-t) \cdot (1-u)$
   $+ o \cdot (1-p) \cdot r \cdot s \cdot (1-t) \cdot (1-u)$
   $+ o \cdot p \cdot (1-r) \cdot s \cdot (1-t) \cdot (1-u)$
   $+ (1-o) \cdot p \cdot r \cdot s \cdot (1-t) \cdot (1-u)$

4. $o \cdot p \cdot r \cdot (1-s) \cdot (1-t) \cdot u$
   $+ o \cdot p \cdot r \cdot (1-s) \cdot t \cdot (1-u)$
   $+ o \cdot p \cdot (1-r) \cdot s \cdot t \cdot (1-u)$
   $+ o \cdot p \cdot (1-r) \cdot s \cdot (1-t) \cdot u$

5. $o \cdot (1-p) \cdot r \cdot (1-s) \cdot t \cdot u$
   $+ o \cdot (1-p) \cdot (1-r) \cdot s \cdot t \cdot u$
   $+ (1-o) \cdot p \cdot r \cdot (1-s) \cdot t \cdot u$
   $+ o \cdot (1-p) \cdot r \cdot s \cdot t \cdot (1-u)$
   $+ (1-o) \cdot p \cdot r \cdot s \cdot t \cdot (1-u)$
   $+ (1-o) \cdot p \cdot r \cdot s \cdot (1-t) \cdot u$
   $+ (1-o) \cdot p \cdot r \cdot s \cdot t \cdot u$
   $+ o \cdot (1-p) \cdot r \cdot s \cdot t \cdot u$
   $+ (1-o) \cdot p \cdot (1-r) \cdot s \cdot t \cdot u$
   $+ o \cdot (1-p) \cdot r \cdot s \cdot (1-t) \cdot u$

6. $(1-o) \cdot (1-p) \cdot r \cdot (1-s) \cdot t \cdot u$
   $+ (1-o) \cdot (1-p) \cdot r \cdot s \cdot t \cdot (1-u)$
   $+ (1-o) \cdot (1-p) \cdot (1-r) \cdot s \cdot t \cdot u$
   $+ (1-o) \cdot (1-p) \cdot r \cdot s \cdot (1-t) \cdot u$
   $+ (1-o) \cdot (1-p) \cdot r \cdot s \cdot t \cdot u$

7. $o \cdot (1-p) \cdot r \cdot (1-s) \cdot (1-t) \cdot u$
   $+ o \cdot (1-p) \cdot (1-r) \cdot s \cdot (1-t) \cdot u$
   $+ (1-o) \cdot p \cdot r \cdot (1-s) \cdot t \cdot (1-u)$
   $+ (1-o) \cdot p \cdot (1-r) \cdot s \cdot t \cdot (1-u)$

8. $o \cdot (1-p) \cdot r \cdot (1-s) \cdot t \cdot (1-u)$
   $+ o \cdot (1-p) \cdot (1-r) \cdot s \cdot t \cdot (1-u)$
   $+ (1-o) \cdot p \cdot r \cdot (1-s) \cdot (1-t) \cdot u$
   $+ (1-o) \cdot p \cdot (1-r) \cdot s \cdot (1-t) \cdot u \cdot$

9. $o \cdot (1-p) \cdot r \cdot (1-s) \cdot (1-t) \cdot (1-u)$
   $+ o \cdot (1-p) \cdot (1-r) \cdot s \cdot (1-t) \cdot (1-u)$
   $+ (1-o) \cdot p \cdot r \cdot (1-s) \cdot (1-t) \cdot (1-u)$
   $+ (1-o) \cdot p \cdot (1-r) \cdot s \cdot (1-t) \cdot (1-u)$

10. $(1-o) \cdot (1-p) \cdot r \cdot (1-s) \cdot t \cdot (1-u)$
    $+ (1-o) \cdot (1-p) \cdot (1-r) \cdot s \cdot t \cdot (1-u)$
    $+ (1-o) \cdot (1-p) \cdot r \cdot (1-s) \cdot (1-t) \cdot u$
    $+ (1-o) \cdot (1-p) \cdot (1-r) \cdot s \cdot (1-t) \cdot u$

11–16.
    $1 -$ (sum of configurations 1–10).

The frequency functions of Table 2.25 are equalled to the frequencies observed for the corresponging configurations, and the six unknowns o, p, r, s, t and u are solved. There are eleven functions, however, and the solution is quite complicated. It can best be done in an iterative way (with the aid of a computer), starting with intelligent guesses for the frequencies of the different segments being bound. It remains complex, but a few examples have been worked out (SYBENGA, 1966c), see Table 2.26.

Table 2.26. The frequencies of the configurations of Fig. 2.24 in the pollen mother cells of the heterozygotes of the reciprocal translocations 248 and 273 of rye (one plant each) and the expected frequencies when the probabilities of o, p, r, s, t, u of being bound for the six segments correspond with those given (cf. Table 2.25). For 248 the equations obtained with Table 2.25 are not solvable (SYBENGA, 1966c, 1970a)

Configuration	Transloc. 248 obs.	exp.	Transloc. 273 obs.	exp.	Segm.	Freq. of being bound used for calculating expect.
1	19	—	5	5		(273)
2	354	—	773	773	O	$o=0.960$
3	467	—	210	208	P	$p=0.985$
4	61	—	1	1	R	$r=0.992$
5	4	—	1	<1	S	$s=0.830$
6	0	—	0	<1	T	$t=0.002$
7	1	—	0	<1	U	$u=0.005$
8	0	—	0	<1		
9	8	—	9	9		
10	0	—	0	<1		
11–16	86	—	1	1		
Total cells	1,000	1,000	1,000	1,000		

The basis of the functions of Table 2.25 is the multiplicability of the frequencies of each separate event, a segment being bound (or not bound). This is rational only when these events are independent for the different segments. Although this may be generally so for segments O and P in respect to each other and to the remaining segments, there are several occasions in which interdependence (interference) has been shown to exist between R, S, T and U. This reduces the validity of the results. On the other hand, as shown for the bivalents (Section 2.1.2), it is possible to use certain (otherwise unexpected) irregularities in the results, such as negative discriminants, as an indication of interference in the translocation complex. When working on the six segments simultaneously such irregularities are not readily detected otherwise than by a failure to find a solution for the equations. This is not enough to pin-point the irregularity as interference between specific segments. For this reason and for reasons of simplicity, a few alterations have been introduced into the analysis, which in many respects make it better accessible (SYBENGA, 1970a).

These simplifications imply primarily a grouping of configuration types into categories in which only the four end segments are involved. The interstitial segments are treated subsequently. Group a comprises all configurations in which the four end segments are bound (1 and 2 in Fig. 2.24). Group b comprises the configurations with three end segments bound (3, 4 and 5 in Fig. 2.24). Group c comprises the configurations with two adjacent end segments bound (7, 8 and 9 in Fig. 2.24), group d those with two opposite end segments bound (6, 11, 12 and 14 in Fig. 2.24) and group e those with one and no end segments bound (10, 13, 15, 16 in Fig. 2.24). Thus the ring-of-four is in group a, but also the "figure-eight", which has interstitial chiasmata in addition to chiasmata in the end segments. Now the first difficulty arises: one of the ways the "figure-eight" is formed is when both interstitial segments have chiasmata but one translocated segment lacks chiasmata. This situation can, at times, be recognized but often this is not the case, especially when the translocated segments are similar in length. Such a configuration, then, does not belong in category a, but in the next: that with one segment devoid of chiasmata (b). The group "figures-eight", therefore, must be partitioned into two classes. This is performed on the basis of the number (one or two) of interstitial segments with chiasmata. Some information on chiasma formation in the interstitial segments is needed, therefore, but not available at this stage of the analysis. All that can be done is to estimate the maximum error caused by misclassification of the "figures-eight". The "figure-eight" must be classified in group b when $both$ interstitial segments have chiasmata, and when one of the two translocated segments lacks chiasmata. Neglecting O and P, which are not relevant for the present purpose, the probability of this event can be given as $(1-r)\cdot s\cdot t\cdot u+r\cdot(1-s)\cdot t\cdot u$ (Fig. 2.25). The two

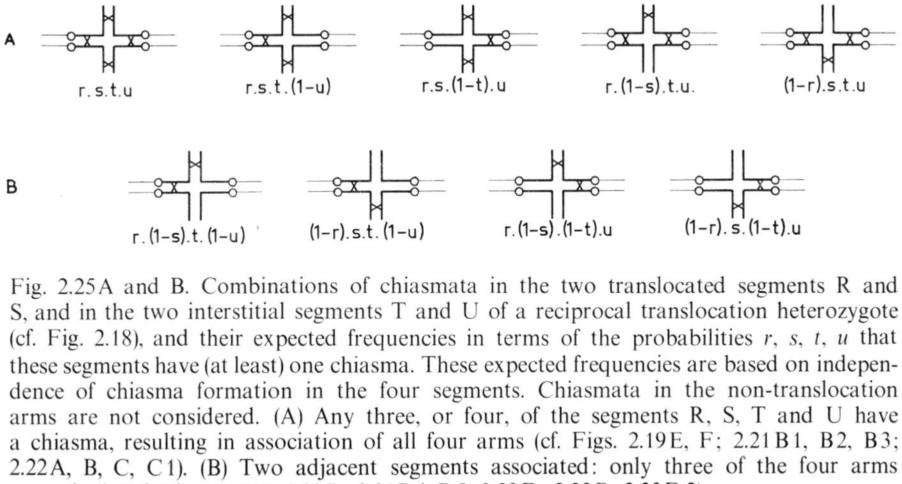

Fig. 2.25A and B. Combinations of chiasmata in the two translocated segments R and S, and in the two interstitial segments T and U of a reciprocal translocation heterozygote (cf. Fig. 2.18), and their expected frequencies in terms of the probabilities r, s, t, u that these segments have (at least) one chiasma. These expected frequencies are based on independence of chiasma formation in the four segments. Chiasmata in the non-translocation arms are not considered. (A) Any three, or four, of the segments R, S, T and U have a chiasma, resulting in association of all four arms (cf. Figs. 2.19E, F; 2.21B1, B2, B3; 2.22A, B, C, C1). (B) Two adjacent segments associated: only three of the four arms are united. (Cf. Figs. 2.18B; 2.20B; 2.21B4, B5; 2.22D; 2.23B; 2.23D2)

terms have $t\cdot u$ in common. This product has its maximal value when $t=u$, the sum being fixed. The total frequency of configurations with interstitial chiasmata (say m) can usually be approximately determined in the multivalent group and gives information on t and u. It equals $m=t\cdot u+(1-t)\cdot u+t\cdot(1-u)$, which for $t=u$ reduces to $m=2t-t^2$, or $t^2-2t+m=0$; $t=\dfrac{2\pm\sqrt{4-4m}}{2}=1-\sqrt{1-m}$.

The further procedures are best demonstrated with an example (cf. Table 2.26). Since in the category "bivalents + univalents" chiasmata in interstitial segments cannot be detected, it cannot be used for the estimate, and the total must be decreased by the number in this category: 86 and 1 respectively in the example. Of the remaining total of 914 cells, 85 had interstitial chiasmata in translocation 248; in translocation 273 there were 7 in a total of 999. It is clear that the 5 "figures-eight" of 273 are not of real significance, and rather arbitrarily 4 are considered as belonging in category a and 1 in b. In 248, however, the 19 "figures-eight" are of somewhat more importance. When $t = u$, we find $t = 1 - \sqrt{1 - \frac{85}{914}} = 0.047$ or $u \cdot t = 0.047^2 = 0.0022$. This is the maximum possible frequency of configurations with both interstitial segments bound. In 914 cells this amounts to 2.01, most of which will have ended up in the "figures-eight". Not much wrong can be done when two of the "figures-eight" are transferred to category b and the remaining 17 to a. This practically neglects: (1) a difference between t and u; (2) the effect of interference (usually negative and therefore acting opposite to a difference between t and u); (3) the possibility that some configurations in which both T and U are bound are not "figures-eight" and finally; (4) the fact that a "figure-eight" still belongs in category a with *both* T and U bound when also R and S are bound. For category a, there now are the totals of 371 and 738 cells respectively in the example, or a frequency of 0.371 for translocation 248 and 0.738 for 273.

Category b, with three out of the four end segments bound is composed of 3, 4 and 5 (Fig. 2.24). The frequency of 5 is too low to cause concern about simultaneous chiasmata in T and U (which would place them into category c). Thus for 248, b has 2 (from 1) + 467 + 61 + 4 cells out of 1,000, or a frequency of 0.534. This is 0.212 for 273.

Category c has two adjacent end segments bound and the other two free. It is composed of 7, 8 and 9 of Fig. 2.24 and has a total frequency of 0.009 for 248 and also 0.009 for 273.

Category d consists of the bivalent pairs, and d' contains the (bivalent + 2 univalents) combinations and the four univalents.

The equations for the frequencies of the four categories expressed in terms of o, p, r and s are as follows:

$$a = o \cdot p \cdot r \cdot s, \tag{2.9}$$

$$b = (1 - o) \cdot p \cdot r \cdot s + o \cdot (1 - p) \cdot r \cdot s + o \cdot p \cdot (1 - r) \cdot s + o \cdot p \cdot r \cdot (1 - s), \tag{2.10}$$

$$\begin{aligned} c = &(1 - o) \cdot p \cdot (1 - r) \cdot s + o \cdot (1 - p) \cdot (1 - r) \cdot s \\ &+ (1 - o) \cdot p \cdot r \cdot (1 - s) + o \cdot (1 - p) \cdot r \cdot (1 - s), \end{aligned} \tag{2.11}$$

$$d = (1 - o) \cdot (1 - p) \cdot r \cdot s + o \cdot p \cdot (1 - r) \cdot (1 - s),$$

$$\begin{aligned} d' = &o \cdot (1 - p) \cdot (1 - r) \cdot (1 - s) + (1 - o) \cdot p \cdot (1 - r) \cdot (1 - s) \\ &+ (1 - o) \cdot (1 - p) \cdot r \cdot (1 - s) + (1 - o) \cdot (1 - p) \cdot (1 - r) \cdot s \\ &+ (1 - o) \cdot (1 - p) \cdot (1 - r) \cdot (1 - s). \end{aligned} \tag{2.12}$$

As expected, there are a total of $2^4 = 16$ terms. In practice, d and d' cannot be separated, since neither can be distinguished from the bivalents and univalents of the chromosomes not involved in the translocation complex. Together, the frequencies of d and d' equal the frequency of cells without multivalents plus the infrequent types 6 and 10 of Fig. 2.24. Since for most translocations configurations of type d' are quite scarce, the results are not seriously affected when all cells without multivalents are taken in group d. We do require five equations, however, in order to make the calculation at all possible: with four equations adding to one, only three degrees of freedom would be available, and that is not enough for estimating four parameters. When group d is large, and there is a suspicion that univalents may occur, it may be realistic to reduce the frequency of d by a small value. An iterative method may be used, first deriving a best estimate for d' from estimates of o, p, r and s made for $d'=0$, and then re-estimating o, p, r and s.

There is one more complication: as already mentioned above, the equations are completely symmetric and as a consequence, although four different values can be extracted, it is not entirely possible to assign these values to specific segments. Two values are found for o and p and two for r and s, but between o and p or between r and s no distinction is possible.

For solving Eqs. (2.9)–(2.12), the first step is to put $o \cdot p = e$ and $r \cdot s = f$; further, $(o+p)=g$ and $(r+s)=h$. Now it can be derived readily that $a=e \times f$ and that $2a+b+d=e+f$. This makes it possible to solve e and f. The value of e, referring to the non-translocated arms, is assumed to be larger than that of f, which refers to the translocated segments. When there are indications of the opposite, the meaning of e and f must be changed accordingly.

$$e = \frac{(2a+b+d)+\sqrt{(2a+b+d)^2-4a}}{2}.$$ (2.13)

The alternative root is f, but the simplest way for finding f, of course, is

$$f = \frac{a}{e}.$$ (2.14)

The derivation of g and h is slightly more complicated. The first step is:

$$4a+2b+c=(o+p)(r+s)=gh \text{ and } 4a+b=f(o+p)+e(r+s)=fg+eh.$$

Then the quadratic equation $fg^2-(4a+b)g+e(4a+2b+c)=0$ can be derived, with root

$$g = \frac{4a+b+\sqrt{(4a+b)^2-4e \cdot f \cdot (4a+2b+c)}}{2f}$$

assuming g to be the larger of the two, unless shown to be the smaller. This can be reduced to

$$g = \frac{4a+b+\sqrt{b^2-4ac}}{2f}$$ (2.15)

and

$$h = \frac{4a+2b+c}{g}.$$ (2.16)

Finally, since $o \cdot p = e$ and $o + p = g$; $r \cdot s = f$ and $r + s = h$, two quadratic equations, each with two roots can be derived:

$$o^2 - go + e = 0 \quad (\text{or } p^2 - gp + e = 0)$$
and
$$r^2 - hr + f = 0 \quad (\text{or } s^2 - hs + f = 0)$$

$$o = \frac{g + \sqrt{g^2 - 4e}}{2}, \tag{2.17}$$

$$p = \frac{g - \sqrt{g^2 - 4e}}{2}, \tag{2.18}$$

$$r = \frac{h + \sqrt{h^2 - 4f}}{2}, \tag{2.19}$$

$$s = \frac{h - \sqrt{h^2 - 4f}}{2}. \tag{2.20}$$

Solution for the examples of Table 2.26 gives some values for o that are slightly higher than 1 (Table 2.27). Reduction to 0.999, when accompanied by a slight increase in p can leave the values for r and s unchanged when very minor adjustments in the "observed" values a, b, c, d and d' are introduced.

Table 2.27. The four groups (a, b, c, d) of configuration types in which only the four end segments of a reciprocal translocation (between two metacentric chromosomes) are considered (cf. Fig. 2.24, Table 2.25) and the corresponding frequencies of these four segments being bound. Since a few values >1 were obtained, slight corrections had to be made (in brackets). Two translocations in rye (1,000 cells each) (SYBENGA, 1970 a)

Translocation	248	273
Configuration		
a	0.371 (0.371)	0.778 (0.778)
b	0.534 (0.533)	0.211 (0.212)
c	0.009 (0.008)	0.009 (0.008)
d	0.086 (0.087)	0.001 (0.001)
Frequency of being bound		
o	1.047 (0.999)	1.011 (0.999)
p	0.943 (0.986)	0.945 (0.950)
r	0.840 (0.840)	0.992 (0.992)
s	0.448 (0.448)	0.822 (0.822)

2.2.4.2.2 Interference

There are two different approaches to the analysis of interference in translocation heterozygotes: one is relatively direct and consists of comparing the relative frequencies of specific configurations which are critical for the detection of simultaneous presence or absence of chiasmata in specific segments. It is primarily of use when the chromosomes are sufficiently different to recognize specific segments (2.2.4.1) but can occasionally also be applied when this is not completely possible. The other approach is indirect and similar to the method used with normal bivalents, where a negative discriminant in the roots of the quadratic equation could be interpreted as an indication of negative interference. This is the approach that can best be applied to the analysis of chiasma formation in the translocations just described, where only a limited number of groups of configurations has been distinguished.

Negative interference between r and s can be detected when it is strong enough to cause the discriminant in Eq. (2.19) and Eq. (2.20) to be negative. Then h, (which equals $r+s$ and thus represents the individual events in R and S) apparently is small relative to f, which equals $r \times s$ and represents simultaneous events in R and S. Superficially it may appear that when in the discriminant the term $(h^2 - 4f)$ is substituted by $(r+s)^2 - 4rs$, the resulting $(r-s)^2$ would necessarily be positive or equal to 0, the latter leading to $r = s$. The quantities h and f, however, are always derived from the frequencies of different configurations, and it is the numerical relation between these configurations that determines the value (which may be negative) of the discriminant. The assumption of independence of different segments, i.e. multiplicability of frequencies, is part of the model and imposes restrictions on the theoretically possible relative frequencies of the different configurations. The frequencies *observed* do not necessarily bother about these restrictions, but if they do not, clearly this particular part of the model is not valid.

Weak negative interference will not generally and positive interference will never result in negative discriminants. When there is negative interference, but not enough to lead to negative discriminants (and thus remains undetected by this method), the estimates of r and s will be closer together than the true probabilities of R and S having chiasmata. This of course is due to the fact that the value of the discriminant is reduced by interference while it is this value which determines the difference between r and s. With positive interference (always resulting in positive discriminants) the estimates for r and s will differ more that the true probabilities. The more closely the actual values of r and s approach each other, the more readily negative interference may be detected; equality of r and s will make their product maximal. At the same time, the chance increases that sampling error causes incidental negative discriminants to occur even when there is no true interference. Large differences between r and s will only permit the detection of negative interference when it is exceptionally strong. It is the same dilemma as discussed in 2.1.1.

Examples of probably significant interference between R and S are given in Table 2.28. Apparently there is variation in the behavior of the same translocation under different conditions. In translocation 248 of rye the two translocated segments usually differ considerably in genetic length. Since nevertheless occasionally the critical discriminants are negative, it is probable that there is occasional interference and that it varies greatly. There are no indications that interference operates between O and P, or between O and P on one hand and R and S on the other. An analysis of these cases is complicated and will not be given here.

Table 2.28. The frequencies of the four groups of configuration types in which only the four end segments are considered for 5 genetically different plants of rye heterozygous for reciprocal translocation 248 and the probabilities of being bound (o, p, r, s) for each of these segments, as far as they could be derived (SYBENGA, 1970b). For a, no solution was possible, due to negative interference between R and S. In b a value >1 was found, requiring a minor correction. In c and f no solvable roots were obtained but with $r=s$ slight adjustments were sufficient (between brackets). Cf. Table 2.25

Plant	a	b		c		d	e	f	
Number of cells	1,000	1,000		500		500	500	500	
Frequency of configuration									
a	0.305	0.371	(0.371)	0.516	(0.512)	0.422	0.204	0.480	(0.480)
b	0.411	0.534	(0.533)	0.400	(0.406)	0.478	0.600	0.422	(0.425)
c	0.007	0.009	(0.008)	0.002	(0.002)	0.000	0.000	0.002	(0.002)
d	0.277	0.086	(0.087)	0.082	(0.080)	0.100	0.196	0.096	(0.093)
Freq. of being bound									
o	—	1.047	(0.999)	—	(0.999)	1.000	1.000	—	(0.999)
p	—	0.943	(0.986)	—	(0.997)	1.000	1.000	—	(0.997)
r	—	0.840	(0.840)	—	(0.717)	0.783	0.728	—	(0.694)
s	—	0.448	(0.448)	—	(0.717)	0.539	0.280	—	(0.694)

The other approach to interference (presence or absence of chiasmata in specific segments as judged form the relative frequencies of specific configurations) is not possible for the two translocated segments R and S when they cannot be recognized. This approach has considerable value for detecting and even estimating positive and negative interference involving the interstitial segments T and U. These segments will be considered next, first in respect to estimating their frequencies of being bound by chiasmata, later in respect to interference.

2.2.4.2.3 Interstitial Segments: Acrocentric and Metacentric Chromosomes

The simplest case for the analysis of crossing-over in interstitial segments in translocation heterozygotes is that of the acrocentric chromosomes. When the centromeres can be recognized (as in metaphase, or when they can be marked with special techniques: c-banding in the mouse, cf. POLANI, 1972; DE BOER, 1974, for instance) association of the proximal (interstitial) regions can be distinguished from association of the translocation segments. In other cases this distinction is not as easy as it may seem. When, in addition, the different chromosomes do not show pronounced size differences, actually all that can be distinguished are configurations very similar to those of Tables 2.27; 2.28, but with O and P replaced by T and U. Raw data are given by LÉONARD (1971) and LÉONARD and DEKNUDT (1967) for the mouse. Estimating crossing-over runs entirely parallel to that in translocations in metacentric chromosomes with neglect of the interstitial segments. It should be noted that if no means are available to distinguish T and U from R and S, it is not possible to assign specific

Table 2.29. The frequencies at diakinesis of configurations in three reciprocal translocations in male mice (acrocentric chromosomes) from which the frequencies of being bound for the interstitial and translocation segments can be derived, using Eqs. (2.13)–(2.20), replacing o and p by t and u respectively. The translocations were induced in the fathers and again studied in the sons; 400 cells studied in each. (LÉONARD and DEKNUDT, 1967)

Translocation	Generation	Configuration frequencies			
		Ring IV	Chain IV	III + I	II + II
T_2 ALD	father	333 (0.832)	28 (0.070)	—	39 (0.098)
	son	333 (0.832)	33 (0.083)	—	34 (0.085)
T_3 ALD	father	232 (0.580)	21 (0.053)	—	147 (0.367)
	son	252 (0.630)	28 (0.070)	1 (0.003)	119 (0.297)
T_4 ALD	father	227 (0.567)	111 (0.277)	1 (0.003)	61 (0.153)
	son	173 (0.432)	137 (0.343)	4 (0.010)	86 (0.215)

frequencies of being bound to the specific segment (Table 2.29). Negative interference again shows up by a failure to solve the roots. Thus, again, crossing-over estimates are confounded with interference. With the more modern methods of locating the centromere by staining the centromeric heterochromatin, much more exact information can be collected (cf. POLANI, 1972; DE BOER, 1974).

In translocations involving metacentric chromosomes, the analysis of crossing-over in the interstitial segments is not excessively complicated when prior knowledge of crossing-over in the exchanged segments and unaltered arms is available (which has been considered in previous sections). Again the cross-over estimates are confounded with interference when specific segments cannot be recognized. The analysis is based either on the frequency of multivalents with interstitial chiasmata (which can usually readily be distinguished from multivalents without interstitial chiasmata, Figs. 2.18; 2.19; 2.20; 2.21; 2.24) or on the frequencies of different forms of bivalents derived from the interchange complex. In the latter case it is necessary to devise methods to distinguish such bivalents from those of the chromosomes not involved in the complex.

Metacentrics: Multivalents with Interstitial Chiasmata. Independent of whether or not O and P (Fig. 2.18) are bound by chiasmata, two groups of multivalents with chiasmata in one or in both of the interstitial segments can be distinguished: those with four arms connected at one point (Fig. 2.25A) and those with three arms connected (Fig. 2.25B). The frequency (i) of the four-arm types can be seen to equal $r \cdot s \cdot t \cdot u + (1-r) \cdot s \cdot t \cdot u + r \cdot (1-s) \cdot t \cdot u + r \cdot s \cdot (1-t) \cdot u + r \cdot s \cdot t \cdot (1-u)$ or

$$i = (t+u) \cdot r \cdot s + t \cdot u \cdot (r+s-3rs).\tag{2.21}$$

The three-arm types have a frequency

$$j = r \cdot (1-s) \cdot t \cdot (1-u) + r \cdot (1-s) \cdot (1-t) \cdot u + (1-r) \cdot s \cdot (1-t) \cdot u$$
$$+ (1-r) \cdot s \cdot t \cdot (1-u)$$

or

$$j = (t+u) \cdot (r+s-2rs) - 2t \cdot u \cdot (r+s-2rs).\tag{2.22}$$

When $(t+u)=k$ and $t \cdot u=l$, we get

$$k = \frac{2\,i(r+s-2\,r\,s)+j(r+s-3\,r\cdot s)}{(r+s-r\cdot s)\cdot(r+s-2\,r\cdot s)}$$

and

$$l = \frac{i\cdot(r+s-2\,r\cdot s)-j\cdot r\cdot s}{(r+s-r\cdot s)\cdot(r+s-2\,r\cdot s)}$$

in which i and j are the (observed) frequencies of the multivalents with four and three arms associated respectively and r and s the (estimated) frequencies of being bound of the two translocation segments, which had been estimated previously. It can now be derived that

$$t = \frac{k+\sqrt{k^2-4l}}{2},\tag{2.23}$$

$$u = \frac{k-\sqrt{k^2-4l}}{2},\tag{2.24}$$

in which it is assumed that $t>u$, but strict allocation of the values t and u to the specific interstitial segments is not possible. Again, negative interference is detected by the discriminant becoming negative: the frequency of simultaneous occurrence of chiasmata in T and U $(l=t\cdot u)$ may not exceed a limit set by independent occurrence $(k=t+u)$. This method of detecting interference has the same limitations as that discussed earlier and again interference is confounded with the estimates of t and u. In addition, any interference between R and S will affect the estimates of r and s and thus in turn affect the estimates of t and u. Table 2.30 gives a few examples of translocations in which enough interstitial chiasmata are available to make the analysis worth while. The statistical significance is difficult to determine. In a number of cases where t approaches u, mere sampling error may cause the discriminant to be negative.

Multivalents with interstitial segments can also be used to make a rough estimate of interference between *adjacent* segments, i.e. between R (and/or S) and T (and/or U). With positive interference between adjacent segments, T and U will have reduced opportunities to form chiasmata when there are chiasmata in R and S. This may be assumed to be more pronounced in the multivalents with four arms associated than in those with three arms associated. With positive interference, therefore, the latter group will be increased relative to the former, or, the ratio i/j will be smaller than expected on the basis of independence. The latter ratio can be derived from Eqs. (2.21) and (2.22), and in a reduced form equals:

$$i/j = \frac{r\cdot s(t+u-t\cdot u)+t\cdot u(r+s-2\,r\cdot s)}{(r+s-2\,r\cdot s)(t+u-2\,t\cdot u)}.\tag{2.25}$$

Table 2.30. Frequencies of translocation configurations with interstitial chiasmata with four arms associated in one point (*i*) and those with three arms associated (*j*), cf. Figs. 2.24 and 2.25. Values *k* and *l* derived from *i*, *j* (this Table) and *r* and *s* (Tables 2.27; 2.28), corresponding with $t+u$ and $t \times u$ respectively. Discriminants in quadratic functions used for solving *t* and *u* (k^2-4l) must be positive, as must be *k* and *l*, for *t* and *u* to be solvable. Expected values for *i* and *j* are those calculated in absence of interference between *t* and *u*. Object: rye; translocations 248, 273 (SYBENGA, 1970b)

Translocation Cells	248[b] 1,000	248[c] 500	248[f] 500	273[d] 1,000
i	0.023	0.036	0.032	0.024
j	0.062	0.054	0.028	0.002
k	0.0702	0.0633	0.0665	0.0389
l	−0.0251	−0.0352	0.0003	0.0128
discriminant	0.1053	0.1449	0.0033	−0.0499
t	—	—	0.0618	—
u	—	—	0.0046	—
i expected	0.0353	0.0503	0.0331	0.0221
j expected	0.0497	0.0397	0.0269	0.0039
signif. of difference	+ +	+	−	−
$t+u \approx i+j$ (multivalents)	0.085	0.090	0.065	—
$t+u \approx$ freq. rings (bivalents)	0.458	0.412	0.323	—

Since in this formula *t* and *u* occur, which have to be derived from *i* and *j*, it is not fit for calculating an "expected" ratio *i*/*j*. A simplification is now necessary. In cases where the total number of interstitial chiasmata is relatively low but still sufficient to yield reliable information, the product $t \cdot u$ becomes small enough to be negligible and Eq. (2.25) may be simplified to

$$i/j = \frac{r \cdot s}{r+s-2r \cdot s} \tag{2.26}$$

depending on *r* and *s* only, for which estimates are available. On the basis of this "expected" ratio, the numbers of expected four-arm associations in the total of multivalents with interstitial chiasmata can be determined and compared with the observed numbers. For translocation 248[b], four- and three-arm associations occurred in 23 and 62 cells respectively among 1,000 cells studied; expected on the basis of the ratio of *i*/*j* were 35.3 and 49.7. The difference is highly significant. Although the discriminant is positive, since *l* is negative, no meaningful roots were obtained. The same translocation in a different background (248[f]), however, did not give a difference between observed and expected: in 500 cells, 16 four-arm and 14 three-arm associations were observed, and 16.5 and 13.5 expected. In the first case (248[b]), roots were obtained for *r* and *s*, in the second (248[f]) negative interference was too strong to derive roots directly, but very minor adjustments in the configuration frequencies led to solvable roots, which, however, naturally were almost equal. As can be seen from Eq. (2.25), the closer r and s approach each other (and consequently, the larger their product $r \cdot s$) the larger the expected ratio i/j. Since this ratio is the criterion for positive interference between the adjacent segments (it decreases when there is positive interference), it is clear that with negative interference between R and S (making the estimates of r and s approach each other), the estimate of positive interference between the adjacent sections will be reduced. Real positive interference then will be stronger than the estimated interference. This may not be forgotten when relating positive interference between adjacent segments to negative interference between opposite segments. With this in mind it appears very probable that, in general, positive interference between adjacent segments is considerable in multivalents in rye.

Bivalent Pairs. When the translocated segments lack chiasmata, the translocation complex is split up into two bivalents provided one or more of the remaining segments do have chiasmata. If they do not, univalents result. The use of bivalents in the analysis of interference is of course only possible when the translocated segments are short enough to allow a relatively large number of bivalents to be formed. Since in most translocations of metacentric chromosomes studied, the frequency of being bound is large for the unaltered arms, for a simplified analysis we assume o and p to approach 1. Then, in cases where R and S happen to be devoid of chiasmata, there are always bivalents and never univalents. There are three categories:

a) Two rings: both interstitial segments have chiasmata; frequency $v = t \cdot u$.

b) One ring and one open bivalent: one interstitial segment has a chiasma, the other does not; frequency $w = (1-t) \cdot u + t \cdot (1-u) = t + u - 2t \cdot u$.

c) Two open bivalents, frequency $x = (1-t) \cdot (1-u)$.

Parallel to what has been derived for the two arms of a normal bivalent, we get

$$t = \frac{2v + w + \sqrt{(2v+w)^2 - 4v}}{2} \tag{2.27}$$

when T is the larger segment, genetically. The other root is u. Again, the discriminant expressed in terms of t and u (for restrictions see 2.1.2 and above in this section) becomes $(t+u)^2 - 4t \cdot u$. When this is negative, there is negative interference.

The difficulty, of course, is to distinguish these bivalents from the non-translocation bivalents. This distinction is possible indirectly when the frequencies of ring and open non-translocation bivalents can be derived from some other, comparable source. This source is the group of cells in which the translocation complex can be recognized, i.e. where it forms a multivalent. It may be assumed that the non-translocation chromosomes behave in the same way in cells with, as in cells without multivalents. Although within-cell between-configuration interactions exist, their net effect is usually limited (BASAK and JAIN, 1963, 1964; SYBENGA, 1967a).

We can classify the cells with a multivalent on the basis of bivalent types: the cells with only ring bivalents (no open bivalents) (frequency A); those with one open bivalent (B); with two (C) etc. The cells without multivalent are classified similarly; but, of course have two more bivalents (or, occasionally, univalent pairs). The classes are here as follows: those with no open bivalents (frequency J); those with one (K); those with two (L) etc. Since pairs of univalents are usually quite rare, they are neglected for the present purpose. When they are not, they must be accounted for. Among the cells without multivalents, A (as in the cells with multivalents) is the proportion in which the non-translocation bivalents are all rings. Now, v, as above, is the proportion of cells in which the two bivalents of the interchange complex are rings. Thus $v \cdot A$ constitutes the class with all bivalents as rings, frequency J:

$$J = v \cdot A . \tag{2.28}$$

This may suffice to estimate v, but usually J is quite small, and the estimate of v thus obtained is poor. There also are the relations

$$K = v \cdot B + w \cdot A , \tag{2.29}$$

$$L = v \cdot C + w \cdot B + x \cdot A , \tag{2.30}$$

$$M = v \cdot D + w \cdot C + x \cdot B , \tag{2.31}$$

etc.

and in addition

$$v + w + x = 1 . \tag{2.32}$$

There are three unknowns (v, w, x) and several equations. A good solution is difficult, but usually only a few classes are sufficiently large, and those are preferred. In the case of more than three large classes, a comparison between solutions obtained from different combinations of equations can be made and the values averaged in some intelligent way. In an experiment in rye (SYBENGA, 1970b) the results were rather erratic, varying greatly when different classes (J, K, L, M) were used for solving v, w and z. In some negative discriminants were found, in others negative values for v, but in a few, acceptable roots for t and u were obtained. Apparently, this approach requires very large numbers of cells without multivalents in a homogenous population, and this is hard to realize.

Interference between translocated and interstitial segments can be studied by comparing the chiasma frequencies in the interstitial segments T and U in bivalents, with that in multivalents. In the case of bivalents, the frequency of T and/or U being bound equals the frequency of ring-bivalents in the translocation complex, since o and p were assumed to be unity. This frequency can be determined simply by subtracting the frequency of rings in the cells with a translocation multivalent from that in cells without a multivalent, correcting if necessary for a difference in average chiasma frequency in the two groups of cells. The difference between the two groups of cells must equal the frequency of ring bivalents formed by the translocation complex. If o and p do not approach 1, a correction must be made. In Table 2.30 these values are compared with $i+j$ from the multivalents, which in that case approximately equals the frequency of T and/or U being bound. It is clear that there is a great excess of interstitial chiasmata in the bivalents (where chiasmata in the translocation segments are absent). In i, of course, in a number of cases both interstitial segments have chiasma, and therefore, in this comparison part of i should be counted double. However, even if i were counted entirely double, the difference with the bivalent rings remains great. This demonstrates a strong, although somewhat variable, positive interference between the interstitial and the translocated segments.

Negative interference between opposite segments is probably more variable and certainly less readily detected, especially between the two interstitial segments. This is at least in part due to the fact that it is detected only when it exceeds a certain limit, which is reached sooner the nearer the segments approach each other in genetic length. In the case of the interstitial segments there is the additional complication that the estimates of crossing over and of interference in multivalents depend on the estimates obtained for the translocated segments. The stronger the interference between the translocated segments, the closer together are the estimates of being bound for these segments and the wider apart the estimates for the interstitial segments. Then the probability of detecting negative interference between the interstitial segments decreases.

It is not unreasonable to assume that negative interference between the translocated segments and between the interstitial segments has its origin in positive interference between the translocation segments on one hand and the interstitial segments on the other hand. The mechanism may be based on pairing difficulties around the break point. Pairing initiated at a number of points along the chromosome proceeds zipper-wise but may not be expected to reach the break point simultaneously from all four directions. When in one segment pairing is complete up to the break point, the adjacent segments may experience considerable torsion which prevents their effective pairing into the direction of the breakpoint (Fig. 2.26). This, then, could be the cause of positive interference between adjacent segments. Since no rigid structure can now be formed around the break point, the remaining segment (opposite that associated first) may have a somewhat better opportunity to associate towards the breakpoint. The segments completely associated and those not completely associated may either be the translocated or the interstitial segments, but there may be a preference. It is a general observation that crossing-over is reduced around the translocation breakpoint (cf. Section 3.1), while some segments may consistently suffer less reduction than others. These are probably the segments that preferentially complete their pairing first.

When variation in pairing is the cause behind the two forms of interference described here, actual presence of chiasmata in the interstitial segments is no prerequisite for negative interference between the translocation segments, and *vice versa*. This explains why translocations with hardly any chiasmata in the interstitial segments still may show negative interference between the translocated segments.

In the system described, where entire segments are studied as a unit, long segments will not show interference very clearly, since reduction of crossing-over near the break point will be compensated by an increase further away in the same segment as a result of normal positive interference. Very short segments

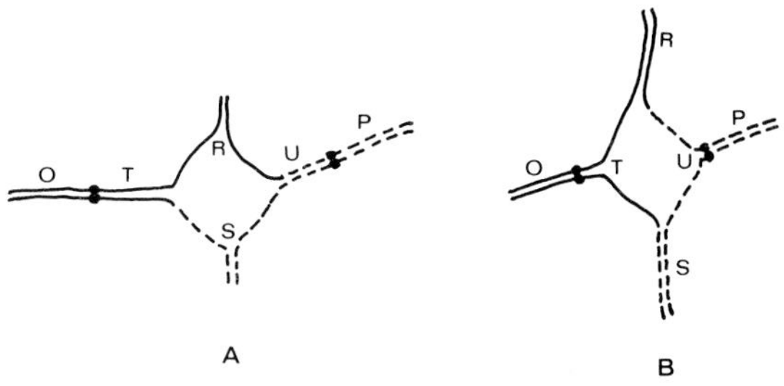

Fig. 2.26A and B. Tentative explanation of positive interference between adjacent segments around the translocation break point and negative interference across the break point. (A) When pairing is completed in T earlier than in R and S, the latter segments will experience difficulties with their lining-up because of torsion, and drag caused by U and P. This may give U an opportunity to pair completely, as it is less seriously affected by torsion as long as R and S are not fully paired. Similarly, when R pairs early, pairing in T and U will be impeded, but S may pair more readily. Thus, whenever pairing around the breakpoint is reduced (which seems to be the rule), the opposite segments tend to behave similarly and the adjacent segments tend to behave differently. The pairing behavior is reflected in chiasma formation

are not bound sufficiently frequently to give reliable information. This probably is not the only reason why SANNOMIYA'S (1968) data do not show any interference around the break point of a reciprocal translocation in *Acrida lata*. The main reason may be that the very short interstitial segment T never paired and left the translocated segments R and S free to move around until they had completed their pairing (Fig. 2.23; Table 2.24).

2.2.4.3 Crossing-over Estimated from Anaphase Segregations

A chiasma in an interstitial segment brings together in one chromosome a chromatid derived from the normal chromosome and one derived from the translocation chromosome. If these differ in length sufficiently, anaphase I chromosomes with visibly different chromatids will appear (Figs. 2.23; 2.27). There are two interstitial segments, and it is convenient when it is possible to distinguish at anaphase I between the chromosomes involved, in order to be able to score for chiasmata in both interstitial segments separately. There are a number of examples (some given in Table 2.23) which were all intended to demonstrate the perfect correlation between chromatid exchange as observed at anaphase I, and the number of chiasmata in the interstitial segments as observed at diakinesis or metaphase I, with the purpose of proving the chiasmatype theory. There is a clear parallel with the experiment of BROWN and ZOHARY (1955) (2.1) with heteromorphic bivalents. These experiments provide interesting examples of estimating crossing-over in the interstitial segments in translocation heterozygotes. In principle, the crossing-over frequency equals the frequency of AI chromatid exchange. A complication arises when more than a single chiasma is formed in one interstitial segment. Whereas in good preparations without much movement of chiasmata two chiasmata may be distinguished at metaphase, the anaphase configurations do not permit the recognition of more than one cross-over event. Two chiasmata, when complementary, result in two translocation chromatids in one chromosome and two normal chromatids in the other. This cannot be distinguished from absence of chiasmata, nor from two reciprocal chiasmata, which cancel each other. Only the two disparate types lead to different chromatids in the same chromosome, a result equal to that of a single chiasma. Only when there is a way to distinguish between the two disjoining chromosomes, can the complementary combination of two chiasmata be recognized, but this is certainly not the rule since except for the translocation, the two disjoining chromosomes are usually homologous. When neither complementary, nor reciprocal chiasmata can be recognized, the frequency of observed chromatid exchange at anaphase I is lower than the frequency of being bound and, therefore, also lower than twice the recombination frequency in the interstitial segment. It is not possible to deduce the crossing-over frequency (or map length) from this chromatid exchange frequency using any existing mapping function. Most interstitial segments are relatively short genetically and will have very few double chiasmata.

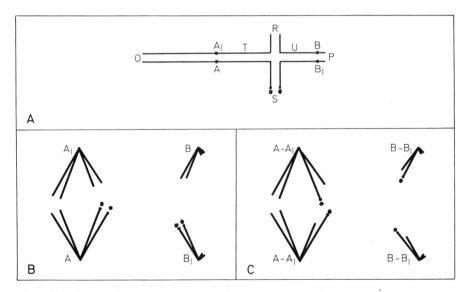

Fig. 2.27A–C. The effect on anaphase configuration shapes of chiasmata in the interstitial segments of a reciprocal translocation heterozygote. (A) The pairing configuration. A and B are the two normal chromosomes (A with a satellite in the long arm); A_1 and B_1 are the translocation chromosomes. (B) without chiasmata in the interstitial segments (or two reciprocal or complementary chiasmata), a normal chromosome A segregates from a translocated A_1, and the same holds for the B and B_1 chromosomes. (C) An interstitial chiasma in T combines a translocated chromatid A_1 with a normal chromatid A. An interstitial chiasma in U does the same for B. (Cf. Fig. 2.21). After JAIN and BASAK, 1963

An interesting example of an anaphase analysis involving more than one chiasma was published by KAYANO (1960). In a reciprocal translocation in *Disporum sessile* one of the translocated segments was heterozygous for a constriction very close to the end. This constriction could be observed at second anaphase. At this stage, the difference between the non-translocated arms at the other end of the chromosome could also be recognized. Thus at AII, recombination between the two chromosome ends could be scored. As can be concluded from Fig. 2.18, only chiasmata in the translocated segments can result in recombination. Since the translocated segment was large and contained up to three chiasmata which could be counted at MI, the material made it possible to study the correlation between multiple chiasma formation and recombination. This is especially important since it permits the detection of chromatid interference, even when the course of the chromatids can not be followed in the chiasmata themselves. As discussed in 2.1.2.1 (compare MATHER, 1938), the frequency of recombination in absence of chromatid interference equals $\frac{1}{2} \times$ the frequency of presence of one or more chiasmata in the segment considered (frequency of chiasmate association or, simply, of being bound). In 1,671 chromatids recovered at AII, KAYANO (l.c.) found 0.251 recombination chromatids. In 1,100 MI cells there was at least one chiasma in 0.512 of the segments considered, which corresponds with a recombination fraction of 0.256. The agreement is strikingly good, indicating that no chromatid interference occurred. Chiasma interference was very low in this material. According to HALDANE'S (1919) adjusted mapping function ($b = 1 - e^{-x}$), which assumes absence of interference, the frequency of chiasmata should have been 0.717 on the basis of 0.512 segments bound. The frequency observed was 0.696. KOSAMBI'S (1944) adjusted function ($y = \text{th}\, x$) gives 0.568, which is expected when interference is average.

KAYANO (1960) also studied chromatid segregation in tetrads and half-tetrads, but since irregular multivalent segregation makes the analysis rather complicated, it is not given here. Again the results closely followed the predictions, which assumed absence of chromatid interference.

Different in respect to set-up as well as results was another analysis by KAYANO (1959), in which AI segregation was studied in relation to chiasma formation in the unequal arm of a heteromorphic bivalent in *Lilium callosum*. A brief discussion of MATHER'S (1935) considerations in respect to chromatid segregation must be given first. In accordance with current terminology, chromatid segregation is termed "reductional" when sister chromatids remain together, and "equational" (as in mitosis) when sister chromatids separate. For most species (including *Lilium callosum*) segregation of the centromere is reductional at AI. For loci beyond the first chiasma segregation is equational at AI, as the sister chromatids are forced to move to opposite poles. Beyond the second chiasma, counting away from the centromere, segregation at AI may be either reductional or equational, depending on the chromatids involved in the two chiasmata (compare Fig. 2.3). Both compensating types of chiasma pairs (complementary and reciprocal) result in reductional segregation, both disparate types in equational segregation. Addition of a third chiasma will make segregation equational when the former two would have resulted in reductional segregation, but half reductional and half equational when the former two would have resulted in equational segregation. "Hence the total proportions of the two types of separation at a locus three chiasmata distant from the spindle attachment will be $\frac{1}{4}$ reductional and $\frac{3}{4}$ equational" (MATHER, 1935). Thus the general formula

$$E_n = \tfrac{2}{3}\left[1 - (-\tfrac{1}{2})^n\right]$$

is developed, in which E is the proportion of equational segregation after n chiasmata. For $n=1$, $E=1$; for $n=2$, $E=\frac{1}{2}$; for $n=3$, $E=\frac{3}{4}$ etc. The graph representing the relation between E and n is wave-like, but with random and even most non-random distributions, in which 0, 1, 2, 3 etc. chiasmata all occur with certain frequencies, E approaches $\frac{2}{3}$ asymptotically with increasing $\bar{n}$. At $n=5$ the deviation from $\frac{2}{3}$ is negligible. With considerable interference, which may restrict the chiasma frequency to almost exactly 1, high equational segregations may be obtained. Whatever the level of chiasma interference, however, the frequency of equational segregation can be predicted directly from the chiasma frequency and distribution on the basis of the above formula, on the condition that there is no chromatid interference. MATHER'S (1935) analysis covered more subjects, but for the present purpose this discussion suffices.

AI chromatid segregation from heteromorphic bivalents with a maximum of one chiasma has been considered before (Fig. 2.13; Table 2.23). KAYANO'S (1959) data on chiasma formation and AI segregation with several chiasmata in the unequal arm are given in Table 2.23C, with the corresponding Poisson distribution (on the basis of the average of 1.962 chiasmata) added. The observed frequency of equational segregation is much higher than expected, and this can only be due to chromatid interference or to sister chromatid exchange. With complete chromatid interference, two chiasmata always give reductional segregation and one and three chiasmata equational segregation. This leads to the expected frequencies of 0.432 and 0.568 respectively (Table 2.23C), close to the observed ones. The range of the frequency of equational segregation (between complete chromatid interference and complete independence) thus is $\frac{2}{3} - 0.432 = 0.235$. The value observed is 0.486, or 0.181 away from $\frac{2}{3}$. Chromatid interference can then be estimated to amount to $\dfrac{0.181}{0.235}$ or 0.77.

In this example, in contrast to the previous one with practically random chiasma formation, the distribution of chiasmata is far from random. On the basis of these two cases, one with both chromatid and chiasma interference, the other lacking both chromatid and chiasma interference, no conclusion in respect to any relation between the two types of interference can be drawn, however.

It may be noted that chromatid interference has been observed by direct observation of the course of chromatids in successive chiasmata (e.g. HEARNE and HUSKINS, 1935, see 2.1.1). It may be a rather common phenomenon, although certainly not universal. If increasing the frequency of complementary chiasmata, it should lead to recombination fractions higher than 0.5, which, however, are only very infrequently reported for higher organisms. Since neither direct (chiasma) nor indirect (chromatid segregation) observations can distinguish between complementary and reciprocal chiasmata, there is no certainty about the type of chromatid interference. If it is merely a shift between the compensating types and the disparate types, there is no effect on recombination. In addition, of course, there is the possibility that the entire effect is due to non-random sister chromatid exchange.

2.2.4.4 Interchromosome Effects Involving Translocations

Heterozygous translocations, like inversions, can cause changes in crossing-over frequencies in other chromosomes in the same cell, as a result of reduced crossing-over in the translocation complex. They have been studied much less extensively than inversions, and for unknown reasons the effect is less straightforward. HINTON (1965) found that in Drosophila, translocations between autosomes reduced crossing-over in the X-chromosome when the interstitial segment was short but increased it when the interstitial segment was long.

The multivalent in meiosis of translocation heterozygotes can also serve as a marker of the complex in a cytological analysis of mutual effects between the chromosomes involved in the translocation and any other recognizable chromosome or complex. In *Delphinium ajacis* ($2n = 16$) BASAK and JAIN (1963) distinguished two sets of chromosomes: two large metacentric pairs and six smaller acrocentric pairs. By introducing reciprocal translocations several new recognizable combinations were made: the introduction of one translocation between two small chromosomes and another between a small and a large chromosome, created four distinguishable groups. In other Delphinium plants a single translocation between a large and a small chromosome made it possible to distinguish three groups. In two more plant types two groups could be recognized, different in structure from each other and from the original karyotype. Chiasmata were counted in numbers of cells ranging from 51 to 150, and subjected to an analysis of variance. In this analysis three components of variance were distinguished: the *internuclear* variance ($n-1$ degrees of freedom); the between groups variance (3, 2 and 1 degree of freedom for the types with 4, 3 and 2 distinct groups) and the between groups *within* nuclei variance with $3(n-1)$, $2(n-1)$ and $1(n-1)$ for the different types respectively.

The latter component is the intranuclear variance, and in all cases was larger than the internuclear variance, highly significantly so in 17 out of 20 plants. This indicates a negative correlation (positive interference) between groups. A further analysis was carried out on specially grouped cells, entirely parallel with that described earlier for between-bivalent correlations (2.1.1.3) and with the same results: with increasing chiasma frequency the variance increased disproportionally, up to a threshold. The authors (BASAK and JAIN, 1963) conclude that competition sets in after a minimum number of chiasmata have formed.

In this example individual chiasmata were counted. A slightly different approach was used by SYBENGA (1967a) in an analysis of interchromosome effects between translocation configurations and the remainder of the genome in *Secale cereale*, rye, $2n = 14$. Two translocations were studied. Merely association of arms was recorded, and only cells in which the translocation complex was present in the form of a multivalent were scored. In the multivalent 2, 3 or

4 arms could be associated, in the five non-translocation bivalents 7, 8, 9 or 10 arms were found associated. From the same floret late and early groups of cells were separately scored, the difference being based on the number of anaphase cells in the group: When many cells were in anaphase, the MI cells of the group studied were considered late; when few or no cells were in anaphase, the group was early. The results are given in Table 2.31. It is clear that no correlation was detected in the early groups (for both translocations) and that a significant negative correlation appeared when the cells approached anaphase.

Table 2.31. Correlations of chiasma-frequencies between two types of configurations (translocation quadrivalents and bivalents) within cells of two translocation heterozygotes of rye. Early vs. late metaphase I cells (SYBENGA, 1967a). In brackets: numbers expected with 0 correlation

Translocation 248

		Early metaphase I Bound arms in 5 bivalents			Totals	Late metaphase I Bound arms in 5 bivalents				Totals
		7	8	9	10		7	8	9	10	
Bound arms	2	0	0	2	3	5	0	0	2	1	3
in multiv.	3	0	18(17)	96(97)	149(149)	263	2	16(21)	108(112)	143(134)	269
	4	0	11(12)	74(73)	112(112)	197	3	19(14)	77(73)	78(87)	177
		0	37	181	274	492	5	35	187	222	449

$$\chi_2^2\,0.319\,(0.90>P>0.80) \qquad \chi_2^2\,5.186\,(0.1>P>0.05)$$
classes < 5 excluded

Translocation 273

		Early metaphase I bound arms in 5 biv.				Totals	Late metaphase I bound arms in 5 biv.				Totals
		7	8	9	10		7	8	9	10	
Bound arms	2	0	0	3	3	6	0	0	1	3	4
in multiv.	3	0	3	39(34)	101(106)	143	0	4	41(64)	109(86)	154
	4	1	20	199(204)	631(626)	851	3	34	153(130)	152(175)	342
		1	23	241		1,000	3	38	195	264	500

$$\chi_1^2=0.975\,(0.50>P>0.30) \qquad \chi_1^2=21.429\,(0.001>P)$$
classes < 5 excluded

As an explanation it was suggested that with anaphase approaching, some associations between chromosome arms are loosened precociously, i.e. anaphase starts earlier for some associations than for others. When it does not start homogeneously about the cell but comes in a wave, some configurations or groups of configurations will have lost several associations before others show any tendency for separation. Usually a lost association cannot at this stage be unequivocally distinguished from non-formation of chiasmata. The wave-wise loss of association produces a negative correlation, which is entirely an artifact. Such an artifact must be accompanied by a lower observed association frequency in the late cells, which was indeed the case in the example considered. The negative correlation in the experiment of BASAK and JAIN (1963) was real: even in the cells with high chiasma frequencies the correlation was strong. It is useful to test the possibility that negative correlations are caused by precocious, wave-wise loss of chiasmata.

2.2.4.5 Telocentric Substitution in Reciprocal Translocations; Crossing-over Estimated from Metaphase Association; Interference

In a telocentric substitution, one of the four chromosomes of a translocation complex is replaced by the two corresponding telocentrics, one for one arm, the other for the other arm. This is possible only for a normally metacentric chromosome (Fig. 2.28). In the best case, the two telos are recognizable at meiotic first metaphase, but more often, depending not only on similarity in size but also on the quality of fixation or on external or genotypic conditions, they will be indistinguishable. There is no fundamental difference with the normal translocation heterozygote, except that now more than 11 of the 64 configurations can be recognized. Many have only a single possible origin and often specific segments can be recognized with certainty. This makes a much more accurate analysis possible than without telocentric substitution. On the other hand there may still be several configurations that are less readily classified, and when these are frequent, especially when in addition the two telos cannot be distinguished, the analysis can become complicated. It may then be useful to introduce a number of simplifications, provided these are not based on unreasonable assumptions. The simplest and often most satisfactory approach is not to work out all possible configurations but to restrict oneself to the types observed, and to reconstruct their origin in terms of chromosomes involved and chiasmata present. Two examples will be given, both from SYBENGA et al. (1973). In one, the two telos could be recognized, very few interstitial chiasmata occurred and the limited number of configurations almost all had a single possible origin. Table 2.32 shows the configurations with the frequencies observed and their most probable origin in terms of chiasmata and chromosomes involved. From the table, the frequencies of being associated can be derived directly for each segment. It should be noted that each segment can be specified. There were five plants. In the original experiment two temperatures were tested for each plant, but since no systematic differences due to temperature were detected, the data for the two temperatures were pooled. The estimates have been collected in Table 2.33. For segment O of plant 70528-10, the frequency o of being bound is derived as $\frac{1,736+163+7+79+12}{2,000}=0.999$; for P, the value $p=0.912$; for R, the value $r=0.996$; and for S, the value $s=0.955$. For the other plants the derivations follow the same lines. There are hardly any chiasmata in the interstitial segments.

It is possible now to analyze interference more meaningfully than previously. The expected frequencies for simultaneous occurrence of chiasmata in R and S can be calculated as the product of their frequencies r and s in Table 2.33.

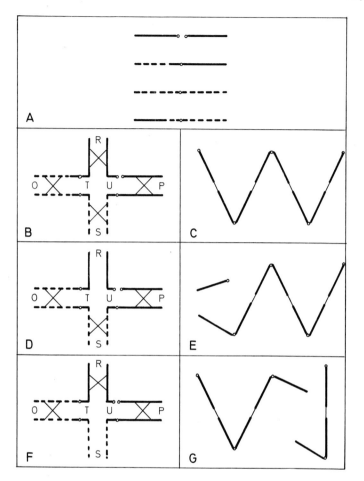

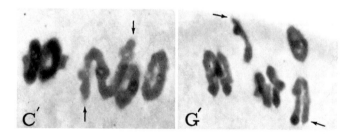

Fig. 2.28 A–G. Telocentric substitution in a reciprocal translocation heterozygote. (A) The chromosomes: one normal metacentric chromosome is replaced by the corresponding two telocentrics. (B), (D), (F) The pairing configuration with chiasmata in different segments. (C), (E), (G) M I configurations corresponding to (B), (D) and (F) respectively; chiasmata completely terminalized; alternate orientation. (C¹), (G¹) Configurations (C) and (G) as observed in rye *(Secale cereale)*. The telocentrics indicated (arrows). In (C¹) the orientation is not alternate, and one normal bivalent is not included

Table 2.32. Metaphase I configurations, their origin and their frequencies in five plants heterozygous for interchange 273, with the non-translocated chromosome VII substituted by two telocentrics, which could be distinguished; 2,000 cells per plant (SYBENGA et al., 1973)

Configurations Pairing M I	Plants 70528-10	70529-13	70530-7	70531-19
\W	1,736	1,816	1,593	1,450
∅/	—	—	—	2
\W˙	163	113	100	121
⌐\/	7	8	12	12
∅/—	—	1	1	—
∨ ∟	79	53	268	383
∟∨	2	2	3	3
∅—∟	—	1	—	1
∀ ∟	—	—	—	2
∨⌐′	—	—	—	1
∨⌐₋	1	3	1	—
∅ ∟⌐	—	—	1	—
>∟∟	—	—	1	—
{∅∧— }∧—	—	1	2	—
∨∧ ᵥ⁻	12	2	18	24
}ᵥ⁼	—	—	—	1
	2,000	2,000	2,000	2,000

The observed frequencies are derived from Table 2.32. For plant 70528-10 the expected frequency is 0.951; the observed frequency $\frac{1,736+163+2}{2,000}=0.951$. The same comparison can be made for the other plants, and the correspondence is striking. This may be partly due to the fact that all values are close to 1, which leaves only little room for variation. The correspondence is too good, however, to make this an acceptable explanation. It is tempting to speculate that in this translocation, with hardly any chiasmata in the interstitial segments, these interstitial segments generally remain unpaired, not in any way restricting pairing in the translocated segments, which thus remain entirely independent. Whether the fact that one of the chromosomes is replaced by two telos, which may move somewhat more freely, also has any effect, cannot be traced.

It may be noted that the genetically longer translocated segment also happens to be the physically longer one.

Another situation prevails in the case of another translocation, also in rye, in which again one of the (normal) chromosomes was replaced by the two corresponding telos. Although the telos were the same as in the previous case, they were not clearly distinguishable in all cells. Several configurations had chiasmata in their interstitial segments. In a number of cases a single metaphase configuration type may have had its origin in more than one combination of chiasmata. In this particular case, prior information on chiasma formation in the different segments was available (cf. Tables 2.26; 2.27; 2.28) and the purpose of the present analysis was to obtain more detailed and unequivocal information on the translocated and interstitial segments. From earlier experiments it was known that O and P both approached 1 closely. For simplifying the present analysis they were assumed to equal 1. It is possible, now, with few exceptions, to isolate those configurations in which R and S are both bound; those in which R is bound, but S is not, and *vice versa*, and finally those in which neither is bound. From these groups the frequencies *r* and *s* are readily determined (Table 2.34). These do not correspond completely with those

Table 2.33. The frequencies of being bound by at least one chiasma for the six segments of a translocation heterozygote with telocentric substitution of one chromosome, derived from Table 2.32. The expected and observed frequencies of simultaneous occurrence of chiasmata in the translocated segments R and S correspond precisely. Four plants of rye, translocation 273 (SYBENGA et al., 1973)

Segment	Frequency	Plants			
		70528-10	70529-13	70530-7	70531-19
O	*o*	0.999	0.999	0.998	0.998
P	*p*	0.912	0.941	0.941	0.927
R	*r*	0.996	0.994	0.991	0.993
S	*s*	0.955	0.972	0.855	0.795
T	*t*	0	0.001	0.001	0.003
U	*u*	0	0	0	0.001
Expected simultaneous chiasmata in R and S: $r \times s =$		0.951	0.966	0.847	0.789
Observed (Table 2.32)		0.951	0.966	0.848	0.788

derived earlier, but this is not unexpected for substitutions. It is interesting to note that the shorter translocated segment has the higher chiasma frequency.

The analysis can be simplified even further by merely comparing the frequency of the quadrivalent—univalent category with that of the bivalent—trivalent category. This gives a superficial but often rather good impression of the relative genetic lengths of the two translocated segments (cf. Fig. 2.28).

Interference between R and S is tested by comparing the expected and observed frequencies expressed in terms of numbers of critical configurations. Of the two plants tested, one did not show any interference, but the other showed highly significant negative interference. Especially in view of this difference it is worth while to analyze chiasma formation in the interstitial segments. For segment T this cannot be done accurately since in several cases neither R nor S are bound, which results in bivalents, and these are not readily distinguished from bivalents formed by chromosomes not involved in the translocation complex. If necessary, an approach comparable to that discussed in Section 2.2.4.2.3 can be tried. Segment U is present in one of the telocentrics and therefore more readily recognized. In plant 70364-15, chiasmata in U were formed in 16 out of the 600 cells, possibly a few more, probably a total of 20. In plant 70364-18 there were at least 32, probably some 40 in 600 cells. Then u equals 0.033 and 0.067 respectively for the two plants. In the category with chiasmata neither in R nor in S the frequencies of chiasmata in U are much higher: 16 in 172 ($=0.093$) and 29 in 144 ($=0.201$) for the two plants respectively. Apparently chiasmata are formed in U almost exclusively when R and S are not bound. This is another clear case of positive interference across the break point between adjacent segments. In plant 70364-18, in which significant negative interference between R and S was observed, the frequency of chiasmata in U was twice that in 70364-15 which did not show interference between R and S. The frequencies r and s were also higher. This suggests a correlation, probably through variation in effective pairing, as also indicated earlier.

Table 2.34. Metaphase I configurations, their origin and frequencies in two plants heterozygous for the reciprocal translocation 248 with the normal chromosome VII substituted by the two telocentrics, which could not be distinguished. A number of configurations have been combined and four groups (A, B, C and D) have been formed according to presence and absence of chiasmata in segments R and S. Cf. Fig. 2.28 (SYBENGA et al., 1973)

Configuration pairing M I	Plant	
	70364-15	70364-18
Group A R and S bound		
	134	198
	0	1
	Total 134	Total 199

Table 2.34 (continued)

Group B
R bound, S not bound

171	169
0	1
0	1
Total 171	Total 171

Group C
S bound, R not bound

122	82
1	0
0	4
Total 123	Total 86

Group D
neither R, nor S bound

156	112
16	29
0	3
Total 172	Total 144

Total	600	600

$$r = \frac{134 + 171}{600} = 0.509 \qquad r = \frac{199 + 171}{600} = 0.617$$

$$s = \frac{134 + 123}{600} = 0.428 \qquad s = \frac{199 + 86}{600} = 0.475$$

	Configurations exp.	obs.		Configurations exp.	obs.
$r \times s = 0.218$	130.8	134	$r \times s = 0.293$	175.8	199
$r \times (1-s) = 0.291$	174.6	171	$r \times (1-s) = 0.324$	194.4	171
$(1-r) \times s = 0.210$	126.0	123	$(1-r) \times s = 0.182$	109.2	86
$(1-r) \times (1-s) = 0.281$	168.6	172	$(1-r) \times (1-s) = 0.201$	102.6	144
	600	600		600	600
$\chi^2_3 = 0.33^{n.s.}$			$\chi^2_3 = 15.711$ $p < 0.01$ negative interference		

2.3 Numerical Markers

Because of seriously reduced viability of hyper- and hypoploid animals, numerical markers are primarily of interest in plants. A few reports are also available for animals, but since the principles are the same, the few examples given have been taken from the more complete botanical material.

2.3.1 Primary Trisomics

In simple primary trisomics, one normal chromosome is present three times, all the others twice. At meiosis, the three homologues form trivalents in a variable proportion of the meiocytes. Since no other chromosomes are involved in trivalent formation, the trivalent marks the particular chromosome. Different combinations of presence and absence of chiasmata in specific segments in the pairing configuration produce characteristic diakinesis and metaphase I configurations. Again, the relative frequencies of such configurations can be used to estimate the frequency of being bound for different segments of the chromosomes involved. There are a few drawbacks in the use of trisomics for estimating crossing-over: since there are three chromosomes which for all practical purposes can only pair in a two-by-two fashion on each particular locus, a variety of pairing configurations is possible, depending on which segment pairs with which homologous segment at any particular point. On this variety of pairing configurations chiasma formation is superimposed. The two systems together produce the observed types of diakinesis-metaphase I configurations: the pairing- and the crossing-over systems are confounded. As long as the pairing is performed according to known

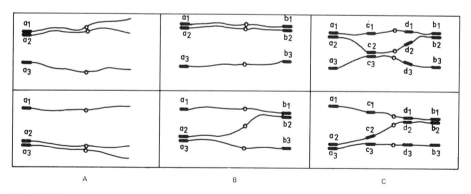

A   B   C

Fig. 2.29A–C. Chromosome pairing in a primary trisomic. (A) With one point (or small region) of pairing initiation (*a*), partner exchange is excluded, and a bivalent and a univalent are formed. (B) With two (independent) points of pairing initiation (*a* and *b*), for instance one at each end, partner exchange is possible but not restricted to any particular site. In absence of partner preference, a 2:1 ratio for trivalent *vs.* bivalent + univalent pairing is expected (Table 2.35). (C) With more points of pairing initiation, partner exchange is more frequent; more trivalents are formed, and more complex pairing and metaphase configurations are possible but their formation is dependent on chiasma frequency and distribution. (SYBENGA, 1972a)

rules (random or not) chiasma formation can be analyzed. Usually, however, these rules are not exactly known *a priori*.

A second drawback is the reduction in chiasma frequency around the point of partner exchange, which makes the results not entirely representative for the disomic situation.

If in the primary trisomic there is only one point from which pairing can start, without exception a bivalent and an univalent will be formed. With two such points, all three chromosomes can associate, with exchange of partners between the two points of pairing initiation. For the sake of simpicity it will be assumed here that each chromosome has two major points of pairing initiation, one at either end (Fig. 2.29). When each has identical opportunities of pairing with both its homologous partners, the association types and resulting pairing configurations of Table 2.35 are formed in the frequencies given. This assumes *random* pairing, but there may be considerable preferences for particular associations, and these will affect the results. In the case of random pairing one third

Table 2.35. The nine combinations of two by two association of the chromosome ends in a primary trisomic. With equal probabilities of pairing, the ratio trivalent: (bivalent + univalent) will be 2:1. (SYBENGA, 1965b)

	paired ends	unpaired ends	pachytene	metaphase I (all arms bound)	
I	$a_1\,a_2$ $b_1\,b_2$	a_3 b_3			bivalent + univalent
2	$a_1\,a_3$ $b_1\,b_2$	a_2 b_3		or	trivalent
3	$a_2\,a_3$ $b_1\,b_2$	a_1 b_3		or	trivalent
4	$a_1\,a_2$ $b_1\,b_3$	a_3 b_2		or	trivalent
5	$a_1\,a_3$ $b_1\,b_3$	a_2 b_2			bivalent + univalent
6	$a_2\,a_3$ $b_1\,b_3$	a_1 b_2		or	trivalent
7	$a_1\,a_2$ $b_2\,b_3$	a_3 b_1		or	trivalent
8	$a_1\,a_3$ $b_2\,b_3$	a_2 b_1		or	trivalent
9	$a_2\,a_3$ $b_2\,b_3$	a_1 b_1			bivalent + univalent

of the pairing configurations consists of a bivalent with an univalent. This bivalent cannot normally be distinguished from other bivalents. The trivalent can be recognized, but whether or not it remains a trivalent up till diakinesis-metaphase I, and what type of trivalent it will be depends on the formation of chiasmata in the different segments. Unlike the case of the translocation heterozygote, where the break point is the fixed position of partner exchange, the point of partner exchange can vary in the trisomic. When in metacentric chromosomes it coincides with the centromere, chiasmata in both arms will lead to a chain trivalent at diakinesis-metaphase, wherever the chiasmata are formed in the arms. When, however, the point of partner exchange occurs somewhere in one of the two arms, there is an interstitial segment, and chiasmata in the interstitial segment together with chiasmata in the terminal segment, and in combination with chiasmata in the other arm, will result in a "frying pan" trivalent (Fig. 2.30). It is clear that in this case both arms are "bound". More than two arms cannot be "bound" at the same time, even though in one arm two different chromosomes are associated with one common partner: of both only a part of an arm is involved. The arm with partner exchange, of course, has two chiasmata, but so may the other arm, where it remains undetected when merely association is counted, and not the frequency of chiasmata (cf. 2.1.2). There may be a third type of trivalent: when partner exchange within an arm is followed by chiasma formation in both the interstitial and the terminal segment, but not accompanied by chiasmata in the other arm, a Y-shaped trivalent is formed (Fig. 2.30). These are three shapes of trivalent resulting from special combinations of chiasmata in the trivalent pairing configuration. With other combinations bivalents result at diakinesis-metaphase I (Fig. 2.30). Such bivalents cannot be distinguished from any other bivalents, and therefore, when the average frequency of chiasmata is not high, the actual frequency of trivalent pairing may be considerably higher than the frequency of trivalents observed at later stages of meiosis.

For an analysis of chiasma formation in the marked chromosome, information on the bivalents formed by this chromosome is indispendable. This can be obtained indirectly in a way comparable to that considered in Section 2.2.4.2.3 which concerned crossing-over in interstitial segments of translocation heterozygotes. It must be postulated that the behavior of the non-marked chromosomes is the same in cells with a recognizable trivalent as in cells without such a trivalent. For a primary trisomic of the satellite chromosome of rye (cf. SYBENGA, 1965b, where a somewhat different approach was used), the data for three different plants are pooled. There were a total of 900 cells, in 351 of which a trivalent was observed. In these cells the not-marked chromosomes were associated as follows:

      1,896 ring bivalents
        206 open bivalents
          4 univalent pairs

Total $\overline{2,106}$ = 6 pairs in each of 351 cells.

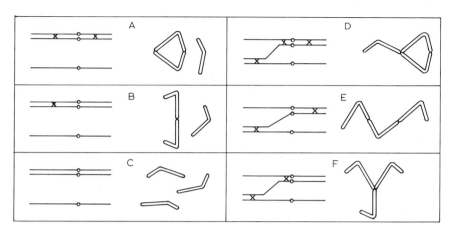

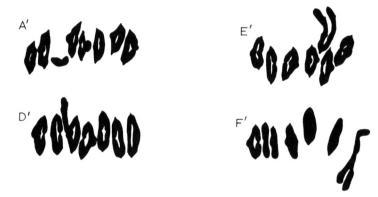

Fig. 2.30 A–F. Pairing and M I configurations (alternate orientation) for bivalent + univalent (A, B, C) and trivalent pairing (one point of partner exchange D, E, F) in a primary trisomic. With one chiasma after trivalent pairing the M I configuration of (B) and without chiasmata that of (C) arise. When in configuration (D) the left most chiasma is absent, the M I configuration of A appears. (A^1, D^1, E^1, F^1) The corresponding configurations in rye (camara lucida drawings). When within a cell the different bivalents can not be identified, a bivalent formed by two of the trisomic chromosomes can not be distinguished from any other bivalent. As soon as open bivalents are formed, no distinction can be made between (A) and (B). (SYBENGA, 1972a)

In the cells without trivalent there were:

> 3,162 ring bivalents
>   660 open bivalents
>   21 univalent pairs

Total $\overline{3,843}$ = 7 pairs in each of 549 cells,

in addition to one univalent per cell. One of the pairs (in combination with the univalent) in these cells is the "marked" chromosome. This pair may be concluded to have formed:

a ring bivalent   in $3,162 - \frac{549}{351} \times 1,896 = 196.7$ cells,
an open bivalent in $660 - \frac{549}{351} \times \quad 206 = 337.8$ cells,
two univalents   in $21 - \frac{549}{351} \times \quad\quad 4 = \;14.6$ cells.

The frequencies of the trivalent types were: chain 295; "frying pan" 31; Y-trivalent 25. As seen from Fig. 2.30, in respect to the number of arms bound, the "frying pan" is equivalent to the chain trivalent, the Y to an open bivalent. The frequencies of the different types (in 900 cells) now are:

trivalents (excluding Y s)	0.362;
ring bivalents	0.219;
open bivalents (including Y s)	0.403;
univalent pairs	0.016.

With a trivalent pairing frequency of f, and consequently a bivalent pairing frequency of $1-f$, the following relations can be set up, in which a is the frequency of the short arm being bound and b the frequency of the long arm being bound:

trivalents:	$f \cdot a \cdot b$,
ring bivalents:	$(1-f) \cdot a \cdot b$,
open bivalents:	$f \cdot (1-a) \cdot b + f \cdot (1-b) \cdot a + (1-f) \cdot (1-a) \cdot b + (1-f) \cdot (1-b) \cdot a$,
univalent pairs:	$f \cdot (1-a) \cdot (1-b) + (1-f) \cdot (1-a) \cdot (1-b)$,

which reduce to:

trivalents	$= 0.362 = f \cdot a \cdot b$,	(2.33)
ring bivalents	$= 0.219 = (1-f) \cdot a \cdot b$,	(2.34)
open bivalents	$= 0.403 = (1-a) \cdot b + (1-b) \cdot a = a + b - 2ab$,	(2.35)
univalent pairs	$= 0.016 = (1-a)(1-b)$,	(2.36)

from Eqs. (2.33) and (2.34) the value of f can be calculated as 0.623, not too far from the expected 0.667 for random pairing. From Eqs. (2.35) and (2.36) a and b can be solved. Via the equation $x_{a,b}^2 - 1.565\, x_{a,b} + 0.581 = 0$ the roots $a = 0.608$ and $b = 0.958$ are obtained. The discriminant is positive.

These are only the best obtainable estimates, and there are several factors that exert an influence on the results, without being accessible to quantitative analysis. For instance, the assumption that the non-marked chromosomes behave the same in cells with a trivalent as in those without cannot be proven. Cases of interchromosomal effects have been reported in rye (cf. Sections 2.1.1 and 2.2.4.1), but may actually be quantitatively negligible in most instances.

Then, sampling error is great in this procedure of subtraction, and will especially affect the frequency of univalent pairs, which is relatively low. Yet it is a crucial category, since it is essential for estimating a and b. As usual in this type of analysis, large numbers of observations are required, and the population from which they are sampled must be homogeneous. A few cells with greatly increased numbers of univalents, for instance, will drastically affect the estimates of a and b and may even cause spurious negative discriminants, which would simulate negative interference.

Third, it is assumed that the frequency of chiasmata is equal in the case of bivalent-univalent pairing as in the case of trivalent pairing. There are indications, however, that in general, partner exchange effects a decrease in chiasma frequency around the point where it occurs. When this is around the centromere, it will not seriously affect chiasma frequencies in organisms with distally localized chiasmata. It will do so in organisms with proximal localization. The point of partner exchange, however, may shift in a trisomic, with a maximum around the middle of the chromosome, which is somewhere in the long arm. There are indeed indications in the example given here, of a difference in chiasma frequency between trivalent and bivalent pairing configurations: when the cells in which all 14 arms of the 7 chromosomes bound are analyzed separately (303 out of 900 cells), there appear to be 174 with a trivalent with both arms bound (chains and "frying pans"). In the 345 cells with one bound arm less (13 arms bound), there are only 113 trivalents with both arms bound. If the loss of bound arms were distributed evenly over the seven pairs of rye, it would happen to a trivalent in $\frac{1}{7} \times \frac{174}{303}$ of the cases, and $\frac{6}{7} \times \frac{174}{303} \times 345 = 169.8$ trivalents would be expected to have remained in the 345 cells (with one bound arm less) instead of 113 observed. The difference is very significant. Furthermore, the average frequency of bound arms in the group of cells with a trivalent equals 0.949 for the not-marked chromosomes and only 0.783 for the marked chromosome. Thirdly, among the cells with all arms bound, a fraction $f = 0.623$ or 561 cells would be expected to have a trivalent for the marked chromosome, but there are onyl 303. These differences can be explained either by an intrinsically low chiasma frequency in the marked chromosome or by reduction in chiasma formation because of trivalent formation. There are numerous indications, which of course, cannot come from the experiment reported here, that there is little difference between the chromosomes of rye in respect to chiasma formation. The conclusion must be, therefore, that trivalent pairing reduces chiasma formation. This implies that the estimate of f, which is derived in fact from the ratio between the frequencies of trivalents with all arms bound and ring bivalents, is estimated too low and may approach the theoretical value of 0.667. A low value for f may have another cause: because the point of partner exchange is not fixed, in a number of cases it may be so near the chromosome end that an interstitial chiasma is formed, but not a distal one. This can result in the formation of a ring bivalent with univalent in spite of the fact that trivalent pairing has occurred. In organisms (like rye) with generally distally localized chiasmata, the frequency of such events will be low.

The combination of information on the trivalent and on the non-trisomic chromosomes is practically non-existent in the rather extensive literature on the meiotic behavior of primary trisomics. Other examples, therefore, cannot be given.

The primary trisomic of an organism with telocentric or pronounced acrocentric chromosomes presents a somewhat simplified case. Trivalent formation is possible only when partner exchange takes place within the arm and chiasmata are formed in the interstitial segment as well as in the distal segment. There is one type of trivalent at diakinesis-metaphase: the Y-shaped trivalent. The restriction that two chiasmata are required in one arm for trivalent detection seriously reduces the possibilities of estimating trivalent pairing frequencies. Even when (by subtraction) the frequencies of bivalents and univalents pairs can be estimated of the marked chromosome in the cells without trivalents, the frequency of trivalents does not contribute any additional information, other than that it is part of the population in which at least two chiasmata occur. It may, of course, have informational value of another character, for instance for possible quantitative segregational consequences of trivalent formation. The subtraction procedure shown to be useful in the case of metacentric chromosomes is much less so in the case of acrocentric chromosomes. The number of types of configurations in the bivalent class is restricted, as there are merely "open", rod-shaped bivalents and univalent pairs, and the latter are quite infrequent in undisturbed material.

2.3.2 Telocentric Trisomics and Monosomics for Normally Metacentric Chromosomes

In the case of the primary trisomic, where a complete chromosome occurs in triplicate, chiasma formation is affected along the entire chromosome and may not be considered representative of the normal disomic situation. It may even be assumed that a reduction by partner exchange may affect one arm more than the other, and in an unpredictable way. With telocentric trisomics, the chromosome is marked equally well, but only one arm is affected by trisomy. In addition, the effect on the phenotype (including the nuclear phenotype) is less severe.

At diakinesis-metaphase again several types of configuration can be distinguished (Fig. 2.31) corresponding with specific chiasma combinations in specific pairing configurations. As in the primary trisomic there are two pairing situations: two chromosomes pair and leave out the third, or all three pair. In the latter case, there is partner exchange, here always within one arm (the trisomic arm) or at the centromere. With complete pairing, if one chromosome is excluded, it is always the telocentric. In the simple case with one point of pairing initiation in each arm, in the vicinity of the end, the trisomic arm has three possibilities of pairing. Two yield a trivalent (telo paired with either one of the normal chromosomes), one yields a bivalent with an univalent (the two normal chromosomes paired). In the trivalent pairing situation chiasmata may or may not be formed in one or more of the different segments, and many more specific configurations can be recognized at diakinesis-metaphase I than in the case of the primary trisomic (Fig. 2.31A). The chain trivalent and the "frying-pan" have both arms bound. Their frequency may be equated to

$$f \cdot a \cdot b, \tag{2.37}$$

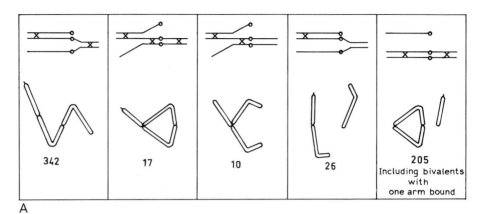

A

B

where f is the trivalent pairing frequency and a and b the frequencies of the two arms respectively being bound by chiasmata. The Y-shaped trivalents and the heteromorphic bivalents also derive from the trivalent pairing situation, their combined frequency being (cf. Fig. 2.31):

$$f \cdot (1-a) \cdot b \tag{2.38}$$

in which b is the frequency of being bound for the trisomic arm. Dividing Eq. (2.37) by Eq. (2.38) gives the value for $\dfrac{a}{1-a}$, from which the frequency of being bound, a, for the not-trisomic arm can be derived directly. This is the arm which is not affected by trisomy and therefore this estimate may be considered representative for the normal disomic situation, provided there is no interference across the centromere. It is independent of f because it is estimated from the trivalent pairing configurations exclusively. An example from SYBENGA (1965c) for the two telocentric trisomics for the satellite chromosome of rye gave the following results: for the long arm telocentric (estimating the frequency of being bound of the short arm), in 600 cells the number of chains trivalents was 342; that of "frying-pans" 17. Thus $f \cdot a \cdot b = 0.598$. The number of Y-trivalents was 10 and that of heteromorphic bivalents 26, giving $f \cdot (1-a) \cdot b = 0.060$. Then $\dfrac{a}{1-a} = 9.967$, and $a = 0.909$. For the short arm telocentric (estimating the long arm) in 700 cells, there were 258 chain trivalents and 23 "frying-pans"; one Y-shaped trivalent and 3 heteromorphic bivalents. Thus $f \cdot a \cdot b = 0.401$ and $f \cdot a \cdot (1-b) = 0.006$. This leads to $\dfrac{b}{1-b} = 66.83$ or $b = 0.985$.

The frequency of being bound of the trisomic arm can be estimated in two ways: an estimate of f can be introduced in Eqs. (2.37) and (2.38), or, the "subtraction procedure" can be followed, similar to that used for the primary trisomics, at the same yielding an estimate for f. In the first case either the "theoretical" value $\frac{2}{3}$ can be taken, or the value obtained from those cells in which all arms are bound. The first is uncertain when preferences for pairing can be expected, the latter suffers from a reduction in chiasma formation in trivalents, which reduces the trivalent frequency in cells with all arms bound. In the example given, in the 600 cells studied of the long-arm telocentric, 287 had all 14 arms bound and among these 211 had a trivalent (necessarily a chain or a "frying-pan", since both arms were bound). This gives a value for f of $\frac{211}{287} = 0.735$, which is considerably more than the expected $\frac{2}{3}$. Apparently,

Fig. 2.31. (A) Pairing and M I configurations of a telocentric trisomic. The chiasma positions leading to the M I configurations shown are indicated. The numbers refer to the frequencies observed in 600 cells of a telocentric trisomic in rye *(Secale cereale)*. With the point of partner exchange located distally, instead of a frying pan, a ring bivalent with univalent may be formed, and instead of a Y shaped trivalent, an open bivalent with univalent. (B) Photomicrographs of the metaphase I configurations in rye; arrows point to the telocentric. In *1*, a chain bivalent; in *2*, a ring (or open) bivalent with telo-univalent; in *3*, a heteromorphic bivalent with metacentric univalent; in *4* a "frying pan" trivalent; in *5* a Y-shaped trivalent. (Cf. SYBENGA, 1965c)

there are pairing preferences and this value is better than the theoretical $\frac{2}{3}$. Introduced into Eqs. (2.37) and (2.38) it leads to a value for the *trisomic b* of 0.895, which is considerably lower than that of the *disomic b* derived from the short-arm trisomic.

Among the 700 cells of the short-arm telocentric there were only 151 with all arms bound and among these a mere 87 with a trivalent. This suggests a value of 0.576 for f, considerably lower than $\frac{2}{3}$. This may again reflect pairing preferences, but in view of the lower average chiasma frequency in this population of cells, it is suspected that a reduction of chiasma frequency due to trisomy may have played a considerable role. In that case this is not a very reliable estimate for f, but it is difficult to find a better one. Applied in Eqs. (2.37) and (2.38) it gives a value of $a = 0.707$ for the trisomic short arm, again considerably lower than the estimate for this arm in disomic condition.

When applying the subtraction method for estimating the frequencies of the different (non-trivalent) configurations formed by the two and a half chromosome of the trisomic complex, an error due to variation in the non-marked chromosomes is introduced. On the other hand, more representative estimates can be obtained than with the use of trivalents only. The example of Table 2.36 is of plant 65261-7 of rye. The frequency of ring bivalents (Fig. 2.31A) can be equated to $(1-f)\cdot a \cdot b$, in which f is the frequency of trivalent pairing. The metaphase I trivalent frequency (including chains and "frying-pans") is $f \cdot a \cdot b$. When the point of partner exchange is close to the end of the chromosome, and no chiasma is formed in the distal segment, trivalent pairing may lead to a ring bivalent. This fraction is small, and neglected here. If not negligible, it may reduce the estimate of f. Open bivalents, when the trisomic arm corresponds with b, have a frequency of $(1-f)\cdot(1-a)\cdot b + (1-f)\cdot(1-b)\cdot a + f\cdot(1-b)\cdot a$. The last term is derived from trivalent pairing. When heteromorphic bivalents and Y-trivalents are included in this category, a term $f\cdot(1-a)\cdot b$ must be added. When a is the frequency of the trisomic arm being bound, the frequency of open bivalents from trivalent pairing is given by $f\cdot(1-a)\cdot b$ and that of the heteromorphic bivalent and Y-trivalent by $f\cdot(1-b)\cdot a$. The univalent frequency (in sets of three) equals $(1-f)\cdot(1-a)\cdot(1-b) + f\cdot(1-a)\cdot(1-b)$.
Or:

ring bivalents with a telocentric univalent:	$(1-f)\cdot a \cdot b$,	(2.39)
trivalents (chain + "frying pan"):	$f \cdot a \cdot b$,	(2.40)
open bivalents + heteromorphic bivalents (both with telocentric univalent) + Y-trivalents:	$a + b - 2ab$,	(2.41)
univalents (sets of 3):	$(1-a)(1-b)$.	(2.42)

The value of f is derived from the ratio between Eqs. (2.39) and (2.40), or for the example of the long-arm trisomic, $f = \dfrac{0.6016}{0.2450} = 0.711$. Although higher than the theoretical $\frac{2}{3}$, it is lower than the value obtained from the frequency of trivalents among cells with all arms bound. This is unexpected, since in general trivalents tend to have reduced crossing-over and as a consequence are reduced to "lower-rank" configurations more readily than ring bivalents. One of the causes for this apparent contradiction may be the formation of ring bivalents after trivalent pairing with a very distal point of partner exchange, as mentioned above. Estimates for a and b are now derived as follows:

$$a+b-2ab=0.0867+0.0167+0.0483=0.1517$$
$$1-a-b+ab=0.0017$$

which results in $a=0.857$ and $b=0.988$.

This deviates considerably from the estimates obtained earlier. Since no direct estimate can be made of the frequency of the trisomic arm being bound, these estimates of a and b depend entirely on the frequency of the class of $2\frac{1}{2}$ univalents, and this consists of only one cell. In order to give an idea of the error introduced by variation in this class, the estimates were made again, now under the assumption that this class contained 2 and 0 cells respectively. The correction necessary to make the sum equal to 1 was made by adjusting the largest single class (chain trivalents). With 2 cells in the univalent class, a and b become 0.866 and 0.975 respectively. With 0 in this class a is 0.8480 and b is 1.000. It is clear that for the trisomic arm, b is consistently larger in the total cell population than in the trivalent containing cell population.

Table 2.36. Application of the "subtraction method" for deriving the estimates for the frequencies of trivalent pairing, f, and the probabilities of being bound a and b, of the two arms of the telocentric trisomics for the long and the short arms respectively of the satellite chromosome of rye. (Cf. Fig. 2.31 and SYBENGA, 1965c)

Bivalent configurations	Long arm telo-trisomic plant 65261-7; 600 cells	Short arm telo-trisomic plant 64413-10; 700 cells
	In cells *with* a trivalent	
6 rings	239	89
5 rings + 1 open biv.	130	137
4 rings + 2 open biv.	22	58
3 rings + 3 open biv.	4	2
2 rings + 4 open biv.	0	1
5 rings + 1 univ. pair	2	1
4 rings + 1 open biv. + 1 univ. pair	1	1
3 rings + 2 open biv. + 1 univ. pair	1	0
1 ring + 4 open biv. + 1 univ. pair	1	0
Total cells	400	289
Total rings	2,202	1,468
open biv.	193	264
univ. pairs	5	2

Table 2.36 (continued)

	In cells *without* a trivalent	
7 rings	79	63
6 rings + 1 open biv.	91	148
5 rings + 2 open biv.	24	140
4 rings + 3 open biv.	3	45
3 rings + 4 open biv.	0	6
2 rings + 5 open biv.	0	1
6 rings + 1 univ. pair	2	2
5 rings + 1 open biv. + 1 univ. pair	1	3
4 rings + 2 open biv. + 1 univ. pair	0	2
3 rings + 3 open biv. + 1 univ. pair	0	1
Total cells	200	411
Total rings	1,248	2,267
open biv.	149	602
univ. pairs	3	8

Derived configuration frequencies of *trisomic chromosome set*

Configuration	Derivation			Derivation		
	(whole) Number		Fre-quency	(whole) Number		Fre-quency
ring biv. + univ. telo	$1{,}248 - \frac{200}{400} \times 2{,}202 = 147$		0.245	$2{,}267 - \frac{411}{289} \times 1{,}468 = 179$		0.256
open biv. + univ. telo	$149 - \frac{200}{400} \times \quad 193 = \ 52$		0.087	$602 - \frac{411}{289} \times \quad 264 = 227$		0.324
univ. pair + univ. telo	$3 - \frac{200}{400} \times \qquad 5 = \ \ 1$		0.001	$8 - \frac{411}{289} \times \qquad 2 = \ \ 5$		0.007
chain trivalent		344	0.574		261	0.373
frying-pan trivalent		17	0.028		23	0.033
Y-shaped trivalent		10	0.017		1	0.001
heteromorphic biv. + univ.		29	0.048		4	0.006
Total		600	1.000		700	1.000

$\dfrac{f}{1-f}$	$\dfrac{0.574 + 0.028}{0.245}$	$\dfrac{0.373 + 0.033}{0.256}$
f	0.711	0.613
a	0.857 (disomic)	0.676 (trisomic)
b	0.988 (trisomic)	0.978 (disomic)

For a comparison, also in Table 2.36, the results for the short-arm telocentric trisomic are given. Since several equations can be constructed, and only three parameters (a, b, f) are to be estimated, numerous solutions are possible. Sometimes it is worthwhile to trace the source of the differences between different solutions.

The telocentric trisomic has been given considerable attention in this Chapter because it permits an analysis of chiasma formation in an undisturbed arm in a chromosome that is well marked. At the same time the behavior both in respect to pairing and chiasma formation in the other arm can be studied, and a comparison between chiasma frequencies in trivalent- and bivalent pairing situations in the same individual can be made.

Telocentric Monosomics. Whereas in diploids the absence of an entire chromosome arm (with rare exceptions) is not tolerated, allopolyploids have frequently been found (or constructed) with one arm less than normal, i.e. one metacentric has been replaced by a telocentric of one of the arms. The number of configurations possible is quite limited: there can be a heteromorphic bivalent between the normal chromosome and the telo, or two univalents. The frequency of the two classes gives a direct estimate of the frequency of being bound of the disomic arm. There are a few examples of such mono-telocentrics, mainly in wheat. Since the primary purpose of the observation was not usually the quantitative analysis of crossing over, the data published tend to be fragmentary. An example has been given in 2.1.1, where it served to demonstrate the correspondence between diakinesis chiasma frequencies and genetic crossing-over (FU and SEARS, 1973).

Chapter 3. The Analysis of Chromosome Pairing

In respect to chromosome pairing, two situations may be distinguished: non-competitive pairing, in which only two possible partners are available, and competitive pairing, where more than two potential partners are available, but where nevertheless only two at a time can finally pair sufficiently effectively to form a chiasma. This introduces an element of choice. The aspects of interest in non-competitive pairing are the number and location of the points of pairing initiation, and the synaptic extent. In competitive pairing, these same aspects are considered in addition to pairing preferences, reflecting pairing differentiation.

Even more than with crossing-over, direct observation is limited. Very infrequently the process of pairing itself can be observed. Occasionally, interesting results have been obtained with quantitative observations of pachytene pairing in organisms with few and small chromosomes in relatively large cells. More often, one has to resort to less accessible stages, such as diakinesis and metaphase I, where association depends on the formation of chiasmata.

Analysis of chromosome pairing is still possible at these stages, but depends on the availability of satisfactory models and often of special constructions. In these models, as can be understood, pairing is usually approached in conjunction with chiasma formation. Several models are discussed and their limitations given. As in the analysis of crossing-over, the special chromosomal constructions serve to mark otherwise not recognizable chromosomes and chromosome segments, and in addition create new conditions. Especially in the study of competitive pairing are such conditions of considerable importance in themselves. Inspection of the Table of Contents makes it clear that competitive pairing has received more attention than non-competitive pairing. This is partly a consequence of the larger number of possible situations, but also of the great theoretical and practical interest of the role of chromosome pairing behavior in speciation, where differentiation is best tested in competition, and in polyploidy.

3.1 Non-competitive Situations

3.1.1 Unmarked Chromosomes

Pairing proceeds in a series of steps: long distance attraction, progression along the chromosome from points of initial contact, and intimate association at the DNA-level as a prerequisite for crossing-over. In non-competitive pairing the first step determines where on the chromosomes the homologues will come in a position to cross-over. It is of importance to know the number of points at which pairing can initiate and where they are located, and what the probability is that pairing initiates at each potential point of pairing initiation. It is generally agreed that pairing proceeds zipper-wise from the point of initiation, but it is also certain that not all points paired this way are available for crossing-over. It has been suggested that the long-distance attraction necessary to bring the homologues together, is the function of a limited number of specific loci: *zygomeres* (SYBENGA, 1966b) in a general way corresponding to the *synaptomeres* of KING (1970) who used this term specifically for the points where the synaptonemal complex starts forming. Functionally, there may be a difference. It is not easy to say when exactly the first step in pairing takes place. In a large number of instances it has been shown (BROWN and STACK, 1968; BROWN, 1972) that premeiotic association is a preliminary to meiotic pairing. Somatic pairing is a normal feature of all cells of Diptera, and can incidentally be observed in many other organisms. It may, in such cases, be a clear parallel alignment of homologous chromosomes, or merely a statistically just significant smaller average distance between homologous than between non-homologous chromosomes. The degree of somatic pairing may be under genetic control and, through differences in premeiotic alignment of homologous chromosomes, affect meiotic pairing as has been suggested particularly for wheat (FELDMAN et al., 1966). If this is a general phenomenon, a quantitative analysis of somatic pairing may be of genetic importance. Such analyses have not only been carried out in wheat and other organisms with incompletely or loosely associated chromosomes, but also, and in much more detail, in the closely associated chromosomes of many gland cells of Diptera, especially the polytene salivary gland chromosomes of *Drosophila* spp. (KELMAN, 1945; SCHULTZ and HUNGERFORD, 1953; JUDD, 1964, 1965). The relation between such association and meiotic pairing is usually not close enough to give the analyses a more than general relevance. Pairing in cells just prior to meiosis is of more genetic interest (e.g. SMITH, 1942; BROWN and STACK, 1968; STACK, 1971) but has usually not been studied quantitatively. It is often pairing between homologous heterochromatin (e.g. MAGUIRE, 1967; CHAUHAN and ABEL, 1968; Table 3.1) either at the chromosome ends or at the centromere. Combined with a general post-anaphase parallel alignment with the centromeres at one side and the chromosome ends at the other side of the nucleus, it provides an effective introduction to meiotic pairing. Even without homologues associating, the parallel alignment, as often observed in bouquet stages where the chromosome ends attach to a certain area of the nuclear membrane, may help in bringing the chromosomes in a favorable position. The actual site of attraction in somatic association may be merely the centromere, as shown by MELLO-SAMPAYO (1973), who found that two complementary telocentrics of wheat, representing the opposite arms of the same chromosome approach each other more closely than arbitrary telocentrics. Such complementary telos only share the centromere. The force keeping them together may last into meiotic metaphase, since coorientations between such telos can be observed to determine anaphase disjunction (STEINITZ-SEARS, 1963). In general, centromere pairing, heterochromatic or not, seems to prevail in somatic association and in non-chiasmate meiotic pairing, in contrast to pairing in chiasmate meiosis, where emphasis is on distal pairing initiation (ROBERTS, 1970, 1972). In *Hylemya antiqua* (the onion fly), VAN HEEMERT (1974) could conclude that in achiasmate males the centromeres were more important in determining pairing, and segregation than the ends. Whether or not this is true also for achiasmate plants (*Fritillaria amabilis*, NODA, 1968) is not known, but the general meiotic behavior of the chromosomes is very similar in both cases. It is well possible that in general in mitosis and achiasmate meiosis centromere pairing is of more importance than in chiasmate meiotic pairing.

Premeiotic alignment of homologous chromosomes is important to prevent interlocking, especially when it occurs at a relatively condensed stage, where entangling is impossible. It is possibly rather universal, and especially striking in the fungus Coprinus, where LU and RAJU (1970) found that the chromosomes synapse immediately after fusion. They were loosened and slightly elongated, but distinct, and rod-shaped. In general the initial contact was in one or both ends, from where the chromosomes "zipped up" into complete and intimate pairing. The centromeres were visible and median or submedian.

In *Lilium longiflorum* "Croft", no premeiotic pairing could be observed (WALTERS, 1970). Fortuitous nucleolar fusion did not cause the nucleolar chromosomes to align. After interphase, but prior to leptotene, the chromosomes contracted and then despiralized again. During the contraction stage no association in pairs could be observed. Although this demonstrates that in this organism somatic pairing evidently is absent, it remains possible that at the end of the contraction stage the homologues have the opportunity to approach each other sufficiently to make efficient zygotene pairing possible. It is always difficult

Table 3.1. Frequency of association in pairs of recognizable homologous chromocentres in *Impatiens balsamina* and *Salvia nemorosa*. The numbers refer to the cells with the indicated numbers of clearly associated pairs. Between brackets the distribution when only the very closely paired homologues are considered (CHAUHAN and ABEL, 1968)

Impatiens balsamina ($2n = 14$)

Anther	Cells analyzed	Numbers of paired chromocenters per cell						
		7	6	5	4	3	2	1
1	50	24(8)	14(13)	2(6)	3(11)	3(3)	3(7)	0(1)
2	30	22(2)	2(7)	2(3)	4(4)	0(8)	0(3)	0(1)
3	30	28(6)	2(6)	0(7)	0(5)	0(3)	0(1)	0(2)
4	30	21(5)	0(5)	7(7)	1(6)	0(5)	0(0)	1(2)

Salvia nemorosa ($2n = 14$)

1	30	19(6)	4(2)	3(3)	1(3)	1(5)	2(6)	0(4)
2	30	17(2)	2(2)	4(7)	1(3)	3(1)	2(2)	1(1)

to prove that no attraction at condensed stages takes place. In general, whenever the strands become visible at leptotene or zygotene, there are already points of contact (MOENS, 1966). Premeiotic attraction of condensed chromosomes finds a parallel in the distributive pairing of condensed, non-homologous chromosomes of corresponding size at meiotic metaphase (GRELL, 1959, 1967).

In somatic pairing, a quantitative analysis is usually not possible, except when the distances between chromosomes are measured at somatic metaphase. Even then, there is no simple correlation with subsequent meiotic behavior.

Meiotic pairing configurations can be recognized for the first time with the formation of the synaptonemal complex. In an analysis by MOENS (1969) the origin could be seen to be close to the nuclear membrane, to which the homologues were attached. At the pairing fork (Fig. 3.1 A, B) thin protein threads could be seen to connect the homologues each of which had a lateral core. The homologues appeared to be pulled together, and when a certain distance between them had been reached, a central element was formed. Then the homologues would not approach each other further. In the synaptonemal complex the two chromatids cannot be distinguished. For reviews of the structure of the synaptonemal complex in various organisms see MOSES (1969) and WESTERGAARD and VON WETTSTEIN (1972). Much has been speculated on the function and structure of the synaptonemal complex. The general, although not universal, attachment of the chromosome ends to the nuclear membrane at the moment of initiation of formation of the synaptonemal complex suggests a relation with the nuclear membrane. At diplotene, the complexes are released and stacked up and in some instances (e.g. *Aedes aegypti*) they can be seen to become modified and associated with various membranes, which again suggests a relationship "by origin or by specialization, between the synaptic structures and nuclear envelope" (FIIL and MOENS, 1973).

Quantitative variation in synaptonemal complex formation in relation to genetic phenomena has hardly been studied, but a few qualitative reports are available. JONES (1973), for instance, reported a difference between the two homologous chromosomes of the grasshopper *Stethophyma grossum* in respect to the width of the lateral element in a segment near the nuclear membrane. One homologue had an enormously enlarged lateral element, and although normal and enlarged lateral elements together formed a central element, it was suggested that the region occupied by the asymmetric synaptonemal complex coincides with the distal region in which crossing-over is normally suppressed in these bivalents. In BOZZINI'S asynaptic *Triticum durum*, no pairing occurred (LA COUR and WELLS, 1970) and no synaptonemal complexes were formed. However, opaque axial cores were present at early prophase, which at later stages were released from the chromatin. The fragments aligned at 625Å distance, even when derived from non-homologous chromosomes. In a number of asynaptic tomato mutants MOENS (1968c) found normal-appearing axial cores at early prophase, and normal but incomplete pairing and incomplete synaptonemal complexes at pachytene. The mechanism of pairing itself seemed to be intact, but possibly the premeiotic alignment had been affected. The total degree of crossing-over was not reduced, but in some mutants the pattern had changed.

Synaptonemal complexes may have an altered appearance in hybrids, without leading to a decrease in chiasma frequency. In a lily hybrid, LA COUR and WELLS (1973) noted an irregularity in the synaptonemal complexes, consisting of breakage and folding of fragments of lateral elements, 0.8 micron long, in heterologous regions immediately after pairing. Central elements were observed in such irregular regions as well as in normal complexes. In an allotriploid

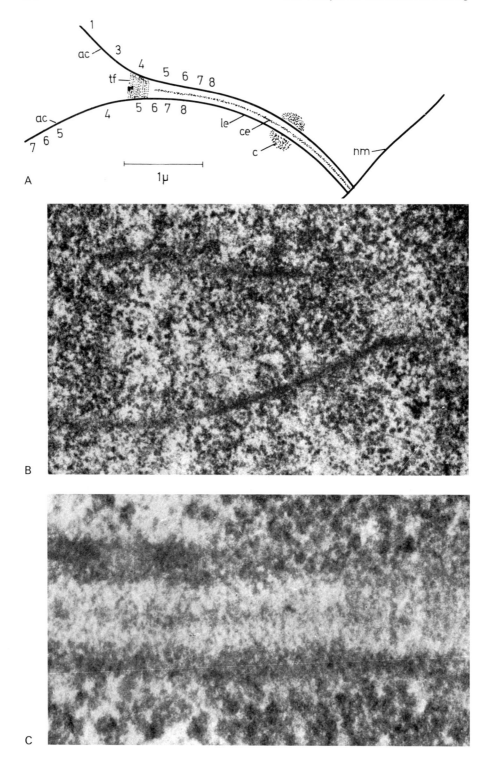

Lilium tigrinum hybrid, MOENS (1968 b) also observed several irregularities (1–2 microns long) in the lateral elements, but only when they were paired. The univalents had normal lateral elements. In their tomato-*Solanum lycopersicoides* hybrid MENZEL and PRICE (1966) found the synaptonemal complex to have only half the width of that in the parents and a greatly reduced chiasma formation. The same authors reported the occurrence of on the average one short stretch of synaptonemal complex in a tomato haploid, in frequency corresponding to the frequency of chiasmata at diakinesis-metaphase. "Although it is difficult to rule out the possibility that the tomato genome might contain one or more small duplicated regions, the variable number and extent of paired regions satisfy the criteria for association of non-homologous chromosomes (McCLIN-TOCK, 1933) as well as is operationally possible". On the other hand, with numerous small duplications only few will find a partner, and these may not be the same every time a synaptonemal complex is observed.

Only in the last case, was a detailed quantitative analysis possible, and a relation between synaptonemal complex characteristics and crossing-over could be demonstrated. A recognizable synaptonemal complex does not seem to be an absolute prerequisite for crossing-over. GRELL et al. (1972) observed that crossing-over in male *Drosophila ananassae* takes place without any demonstrable synaptonemal complex. The level of crossing-over, however, was considerably lower than in the female which clearly has one.

In a number of organisms, especially those with short chromosomes in small numbers, in light microscope squash preparations of pachytene, pairing can be analyzed in more detail. Then the *final* stage is studied and not the process of pairing. Only a few examples of the many analyses can be given. Pachytene pairing has been studied in several haploids. Usually extensive pairing is observed. Some of this must be homologous pairing, due to duplication in the genome, but much of it is non-homologous, as generally appears from the parallel alignment of stretches of chromosome with clearly different chrommere patterns. It may even be seen to initiate at non-homologous loci, for instance in the heterochromatic terminal segments in rye haploids (LEVAN, 1942). Such non-homologous pairing initiation would decrease the chance of homologous segments finding each other. In some cases bivalent pairing prevails (LEVAN, 1942, rye) in others, partner

---

Fig. 3.1 A–C. Development of the synaptonemal complex: (A) Diagram: pairing has started at the chromosome ends, often (as here) associated with the nuclear membrane (*nm*). In the completely paired segment the two axial cores (one for each chromosome) are aligned at a certain distance from each other, forming the *lateral elements* (*le*) of the synaptonemal complex. The formation of the *central element* (*ce*) during pairing of the axial cores (*ac*) is preceded by the formation of protein strands (transverse filaments, *tf*) connecting the two chromosomes. The diagram is constructed from a series of consecutive electron micrographs of *Locusta migratoria* spermatocytes. One of the series is shown in (B), where the axial cores are seen to approach each other; some of the transverse filaments are visible. (C) Completed synaptonemal complex at zygotene-pachytene in *Lilium tigrinum*. Note the lateral and central elements and the DNA/protein filaments traversing the space between them. The dark material outside the synaptonemal complex is the chromatin of the paired chromosomes. (Courtesy P. B. MOENS; cf. SYBENGA, 1972a)

exchange may be very frequent (RIEGER, 1957, Antirrhinum). In again other instances haploid pairing is very limited, as in *Nicotiana tabacum*. When in such cases duplications of known origin occur, pairing can be readily studied in the duplicated segment, although it may not be inferred that the pairing of this segment is quantitatively identical to that in the diploid. An example is the haploid "Coral" tabacum, analyzed by LAMMERTS (1934). The duplication was derived from a translocation, and of considerable length. The configurations observed were more complex than expected on the basis of pairing in the duplication alone. It is possible that in tobacco in general non-homologous pairing is not readily initiated, but once it starts in two homologous segments, can continue into non-homologous areas. In addition, however, pairing would initiate in numerous, otherwise apparently unaffected chromosomes, but limited to shorter chromosome segments (Table 3.2; Fig. 3.2). In general, very little quantitative information on haploid pairing is available (cf. MAGOON and KHANNA, 1963; KIMBER and RILEY, 1963).

Of course in the *polyhaploids* of polyploids, extensive homologous or homeologous pairing may take place. Homeologous pairing in such cases is equivalent to pairing in species hybrids. It is under genetic control and can be effective even when the morphology of the homeologous chromosomes at other than

Table 3.2A–C. Pachytene pairing (A), metaphase bivalent frequency (B) and metaphase fragment (from crossing-over in assymetrically paired chromosomes) frequency (C) in a haploid of "coral" tobacco carrying a duplication, and in a control haploid of comparable origin (LAMMERTS, 1934). For both: $n = 24$

(A)	Number of paired threads per cell at pachytene									
	0	1	2	3	4	5	6	7	Many	Total cells
"Coral"	3	17	14	11	8	2	2	2	4	63
Control	83	22	8	3	0	0	0	0	0	116

(B)	Number of bivalents per cell at metaphase I						Average per cell	Total cells
	0	1	2	3	4	5		
"Coral" (greenhouse)	54	94	81	27	8	3	1.44	267
Control (greenhouse)	265	83	31	8	0	0	0.43	387
Control (field)	847	113	31	4	0	0	0.17	1,095

(C)	Number of fragments per cell at metaphase I			
	1	2	3	4
"Coral" (greenhouse)	78	35	8	1
Control (greenhouse)	25	3	1	0
Control (field)	41	0	0	0

pachytene stages is rather different (BROWN, 1958). In the tetraploid, when identical homologues are available, pairing tends to be preferentially between these and to be scarce between homeologues, but the degree of preference again depends on the genotype.

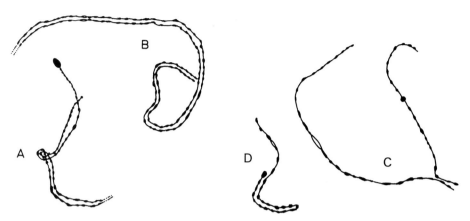

Fig. 3.2 A–D. Pachytene associations in *Nicotiana tabacum* haploids. (A) Typical non-homologous, incomplete association in the coral haploid. (B) A long segment right of the centromere shows homologous pairing, and probably represents the coral duplication. Left of the centromere, pairing apparently is non-homologous: the paired chromomeres do not match. (C) Typical associations in normal haploids: only short segments are paired. (D) Foldback (non-homologous) as often found in the coral haploid. Copied from camara lucida drawings by LAMMERTS (1934)

3.1.2 The Relation between Cross-over Frequency and Synaptic Extent

In LAMMERTS' (1934), haploid "coral" tobacco with a duplication, recombination (scored as the number of bivalents, each with one chiasma) was compared with extent of pairing at pachytene (Table 3.2). Although both were recorded rather crudely, there is a good correspondence. The increase in crossing-over is primarily in the class with one bivalent and apparently due to the duplication. The other classes (two bivalents and higher) had also significantly increased in number. On the assumption that crossing-over is possible only between homeologously paired chromosome segments, it must be concluded that the increase in pachytene pairing, although often between structurally visibly different chromosomes, involved a high degree of homeologous pairing, either between duplicated segments or between homeologous chromosomes (tobacco is an allotetraploid). The conclusion must be either that homologous stretches alternated with non-homologous stretches, or that some homeologous segments have a different chromomere pattern. The first alternative seems reasonable in this case since the frequency of fragments resulting from crossing-over in partly homologous chromosomes, from "fold backs" or assymetric duplications, also increased strikingly (Table 3.2). At metaphase, the duplication bivalent could not unequivocally be distinguished from other bivalents. Fold-backs representing homologous pairing must result from duplications within a single chromosome, but bivalent pairing results from duplications in other chromosomes. These may be *within* a genome, or they may simply reflect the homeology *between* the two genomes in the (di)-haploid. If, in the latter case, homeologous regions alternate with non-homologous regions, the two genomes must have experienced considerable structural differentiation since their first separation from a common ancestor. Such an extensive differentiation seems improbable, and therefore, a considerable within-genome duplication may be the reason for the high haploid bivalent frequency. Superficially it would seem unexpected that pairing between duplications within the genome is relatively more frequent than pairing between homeologous genomes, as must be concluded from the foregoing. The explanation may be that the chromosomes of the two parental genomes of tobacco had differentiated so much in essential pairing characteristics prior to being combined in the allopolyploid, that although in gene content quite comparable, pairing between them has become almost impossible. Within the genomes, the essential pairing characteristics (timing of different processes, for instance) are more similar, and when there is a potential point of pairing initiation in homologous segments, there is a chance for pairing when complete homologues with optimal pairing conditions are absent.

In tobacco haploids with limited non-homologous pachytene association, a comparison between pairing and crossing-over (bivalent formation) can be made. In other haploids with extensive non-homologous pairing, this is not meaningful. Then diakinesis and metaphase chiasmata can be used as an indication that effective homologous pairing has taken place. If it may be assumed that chiasmata are formed only between homologous segments, the frequency and distribution of chiasmata give information on the number of duplications present. The *variation* in frequency and distribution of chiasmata gives information on the frequencies of (effective) pairing of the duplicated segments, and therefore is of more importance for the present context than frequency and distribution as such.

The relationship of cross-over frequency to synaptic extent at pachytene has been studied extensively by MAGUIRE in maize, using chromosomal rearrangements of various types, in competitive as well as non-competitive pairing situations. Her primary interest in these studies was to find out "whether chromosomes tend to (or must) complete observable synapsis before crossing-over occurs or whether it can immediately follow the homologous meeting of short segments" (MAGUIRE, 1965). This was tested by comparing the frequency of pachytene synapsis of a relatively short segment with the frequency of formation of a chiasma, observed at metaphase I. This required a chromosomal marker. In one experiment a segment was marked by trisomy. A small extra chromosome derived from Tripsacum was present in addition to two chromosomes of larger size, which were either entirely or for the greater part made up of maize chromosome 2. In the latter case a stretch of Tripsacum chromosome replaced a segment of maize chromosome. Whereas the two long chromosomes were homologous over the majority of their length, and always paired and formed chiasmata, the short chromosome was homologous with only a restricted segment of one or of both of the long chromosomes. Trivalents were formed in a number of cells and could be recognized at pachytene. The length of the segment of the short extra chromosome actually paired could also be determined (Table 3.3). The trivalent frequency as such is not of interest in the present context. What matters is the number of trivalents that were recovered at metaphase (equalling the number of times a chiasma is formed in the segment considered) compared with the frequency (and length) of association at pachytene. In all combinations of the extra chromosome with normal or derived chromosomes 2, there was excellent correspondence between the frequency of pachytene association and chiasma frequency at metaphase (Table 3.3), even though the length of the paired segment varied. The maximum genetic length available for pairing in the extra chromosome was 29 map units, corresponding with a theoretical maximum of 58 percent chiasmata after pairing. In the best analyzed plant, the actual metaphase chiasma frequency was 43 percent, but since only 50 percent had paired at pachytene, the chiasma frequency in *paired* segments was 86 percent.

In another experiment (MAGUIRE, 1966) an inversion of a length corresponding to 19 genetic map units was studied in the heterozygote. The percent reverse synapsis (loop) was compared with the percent of crossing-over in the inversion segment, scored as anaphase bridge frequency (cf. Figs. 2.14, 2.15, 2.16, 3.3 A, B). The percent cells with a fragment only (also containing a chromatid loop which is difficult to detect in maize) was added to the percent cells with a bridge and a fragment (Table 3.4). Although frequently the length of homologous synapsis was only a small proportion of the total inverted region, the correspondence between frequency of synapsis and frequency of chiasma formation is striking. This correspondence led MAGUIRE (1965, 1966) to conclude that either crossing-

Table 3.3. The relation between average % paired pachytene length of the short extra chromosome, the average % pachytene trivalent frequency and the average % metaphase I trivalent frequency in maize trisomics. The extra chromosome was a tripsacum derivative of varying constitution and for various lengths homologous with one or two corresponding maize chromosome derivatives (MAGUIRE, 1965). Numbers of cells 100—appr. 1,500 per entry

Constitution	Av. % length paired at pach.	Av. pach. trivalent frequency	Av. metaphase trivalent frequency
a	20	98	90
b	15	65	70
c	20	99	93
d	15	17	15
e	11	50	48
f	11	18	17
g	23	60	67
h	16	64	71
i	20	66	72
j	23	100	95
k	16	54	58

corr. coeff. pach.-metaphase trivalent frequency + 0.97; regression 0.99

over always follows synapsis, independent of the size of the synapsed segment, or that the site of crossing-over is the primary contact site from which zygotene pairing proceeds.

The relationship between the pattern of pachytene pairing and the amount of genetic crossing-over in maize was also studied by RHOADES (1968), who

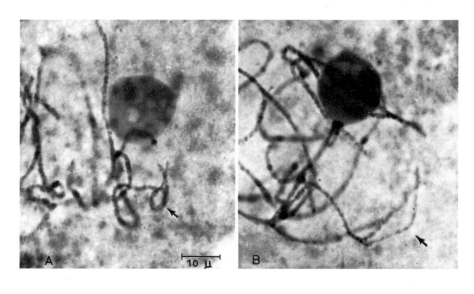

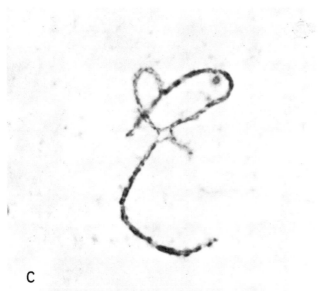

Fig. 3.3 A–C. Pachytene pairing in structural heterozygotes of *Zea mays*. (A) Fairly complete loop pairing in an inversion heterozygote. (B) No pairing in the same segment (MAGUIRE, 1966). (C) Cross-configuration of a reciprocal translocation heterozygote. Note pairing failure around the break point. (Courtesy M. M. RHOADES)

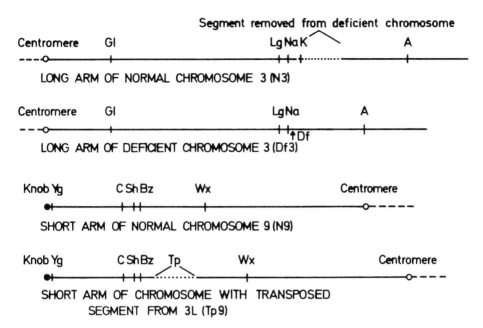

Fig. 3.4. Transposition (interstitial translocation) of a small segment of chromosome *3* to chromosome *9* of *Zea mays*. (RHOADES, 1968)

used a small transposition between the long arm of chromosome 3 and the short arm of chromosome 9 (Fig. 3.4). The transposed piece was approximately 10% of the length of the long arm of 3. Studies on pachytene synapsis were combined with genetic recombination analyses. In the transposition heterozygote a buckle was expected to be formed of the N3 chromosome in the Df3 N3 bivalent at pachytene, and of the TP9 chromosome in the TP9 N9 bivalent. The transposed segment was too small to cause any appreciable multivalent formation. There was a knob present in the long arm of the normal chromosome 3, which served as a reference site in delimiting the location of the buckle in Df3 N3 bivalents. The buckle consistently included the knob, and there was little variation in the relative length of the buckle. In genetic studies involving

Table 3.4. Relation between pachytene pairing and anaphase bridge and fragment formation for a paracentric inversion heterozygote in maize (MAGUIRE, 1966)

Plant	Pachytene					Anaphase I			
	% non-homol. pairing	% no pairing	% loop pairing	Cells	Unclassi-fiable	% normal	% bridge + fragm.	% fragm. only	% sum
1	46.5	17.4	36.0	669	164	64.2	20.9	14.9	35.8
2	57.0	12.9	30.1	667	172	70.4	17.8	11.8	29.6
3	35.3	29.8	34.9	671	127	65.8	18.6	15.6	34.2

the markers *Lg*, *A* and *Na*, which flank the deleted segment, a marked reduction in recombination was observed. Compensating increases occurred in adjacent regions.

In Tp9 N9 bivalents there was a general reduction in crossing-over for all regions in 9 S. At pachytene, the buckle occurred at various positions, proximal, interstitial or distal and was variable in length. In many bivalents no buckle at all had been formed, and despite one short arm being longer than the other, they associated troughout their lengths.

This difference between chromosomes 3 and 9 was not entirely reflected in the behavior of the transposition homozygote in respect to crossing-over. Pachytene pairing was complete in both cases. No difference in the amount of recombination (17%) between *Sh* and *Wx* in Tp9 and N9 homozygotes was found, even though in the former the physical length of this region was extended by the inserted piece of 3 L, corresponding to 20% of the short arm of N9. In the more distally located segment between *Sh* and *C*, crossing-over was doubled, apparently as a result of a lapse of interference between *Sh* and *Wx*. The great majority of the exchanges in *Sh–Wx* occur at the other side of the insertion, and although the segment derived from chromosome 3 is free of crossing-over, its presence increases the distance between the originally adjacent regions sufficiently to make crossing-over in the two regions practically independent of each other. In the Df3 chromosome, the markers *Lg* and *A* lie to the left and right of the deficiency and although separated by considerably less chromatin than in normal chromosomes 3, have an equal amount of recombination (31.1 *vs.* 30.6 percent). Evidently, the transposed segment is free of recombination, independent of its position. It is not genetically inert, since N9 Df3 spores abort. It is noteworthy that in the presence of B chromosomes, which had no effect on crossing-over in the *C-Wx* region in normal chromosomes (cf. Fig. 3.4), recombination in the Tp9 homozygotes was more than doubled in this same region containing the L3 segment. The intimacy of pairing is not discernibly altered. In normal chromosomes 3, recombination between *Lg* and *A*, in which region the same segment occurs, (see above) is not increased in the presence of B chromosomes. It is clear that the correlation between pachytene pairing and crossing-over is not always as straight-forward as suggested by MAGUIRE (1965, 1966). This appears also from EINSET'S (1943) observations on maize trisomics: pachytene trivalent frequencies were generally higher than metaphase trivalent frequencies. This report has the complication that diplotene trivalent frequencies were of the same level as those for pachytene and that diakinesis frequencies were close to those for metaphase. It is difficult to choose between the possibility that the diplotene estimates are correct (which implies that chiasmata had disappeared between diplotene and diakinesis) and the possibility that the diplotene scores were incorrect in the sense that the associations observed were not chiasmate but merely a relict of pachytene pairing. If the first is true, there is a close relationship between pachytene pairing and chiasma formation. If the second is correct, there is no close correlation.

The effect of variation in pairing (studied at pachytene) on recombination had been analyzed already in 1934 by BURNHAM for maize interchanges. In the long arm of chromosome 9, the genes *Yg₂*, *Sh*, and *Wx* lie in this order.

Cross-over percentages were given as 23 for Yg_2–Sh and 20 for Sh–Wx. In an interchange between the *short* arm of chromosomes 9 and a small piece of the long arm of 5, the pachytene cross varied greatly in position, in spite of a usually close association of the chromosomes. Much of the pairing must have been non-homologous. Although the translocation was not in the arm of the genes studied, recombination dropped to 11 for the more distal segment (Yg_2–Sh) and 6 for the more proximal (Sh–Wx) segment.

In a later study of a translocation between chromosomes 1 and 7, again in maize, BURNHAM (1948) again reported variable asynapsis around the break point, cf. Fig. 3.3C, with a shift in the position of the pairing cross. In the long arm of 7, the apparent length of the interchanged piece varied between 67.3% and 91.8% (average 82.6 ± 5.6) of the arm. Asynapsis was common: in 14 partially asynaptic cells it ranged from 10.0 to 35.3% of the long arm. In four cells synapsis was complete. The average asynapsed length was 14.4%. The gene Gl was located 2.9% cross-over units from the translocation, and Ij further away, normally 20% from Gl. The reduction of the latter distance in the heterozygote was down to 8.3% on the female side and 7.6% on the male side.

3.1.3 The Location of Pairing Initiation

Reduction of crossing-over in the vicinity of a translocation break point has been used by ROBERTS (1970, 1972) to locate the major regions of pairing initiation in *Drosophila melanogaster*. Although pachytene pairing is not accessible to analysis in Drosophila females, the location of the break points can be accurately determined in polytene chromosomes, and many markers are available to study changes in crossing-over. Translocations tended to be most effective as cross-over reducers when one of the breakpoints was near (not at) the tip in some instances, and in other instances when positioned at about $\frac{1}{3}$ from the tip. Reductions from 44% crossing-over to less than 5% were observed. There was considerable variation between chromosomes, and the second break point was also of importance. The effect was shown not to be due to elimination of cross-over chromatids. The maximum effect was not reached in segments with maximal crossing-over, but generally nearer to the tip. It was suggested that not necessarily the tip itself was the major point of pairing initiation, but a site slightly removed. The tips probably associate with the nuclear membrane, but not at specific sites. At a short distance from the membrane the homologues can find each other and pair (cf. MOENS, 1966, 1969). It is possible that a second region of pairing initiation occurs nearer to the middle of the chromosome arms, but the proximal regions were ineffective in respect to pairing initiation.

An intricate analysis of chromosome pairing initiation with the use of translocations, not based on alterations in crossing-over but on alterations in pachytene pairing was carried out by BURNHAM et al. (1972) in maize. A series of reciprocal translocations between chromosomes 1 and 5 were used, and four different types were distinguished, according to the arm involved. In one type (five translocations) the short arm of 1 and the short arm of 5 were involved (SS); in the second (five translocations) the long arm of 1 and the long arm of 5 (LL); in the third (six translocations) the short arm of 1 and the long arm of 5 (SL) and in the fourth (eight translocations) the long arm of 1 and the short arm of 5 (LS). The break points were determined as accurately as possible, which in view of the variability of the position of the pachytene cross was not always easy, and some positions were somewhat uncertain. Different combinations of translocations were made and pachytene pairing analyzed. Fig. 3.5 gives an example of the chromosomes involved in a SL × LS intercross, and the expected pattern of pairing. With complete pairing, a double pachytene cross is expected. However, when pairing involves only the interstitial segments, two bivalents will result. Two bivalents are also formed when only the homologous

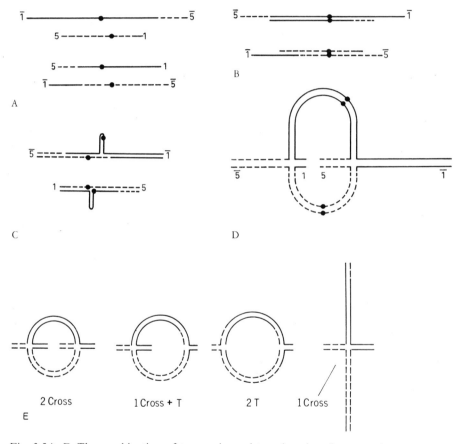

Fig. 3.5A–E. The combination of two reciprocal translocations between chromosomes 1 (drawn in full line) and 5 (dashed) of *Zea mays*. Of each chromosome one end is marked by a dash over the number. (A) The four chromosomes. (B) Homologous pairing of the central portions; more distal pairing is non-homologous, ends unpaired. (C) Homologous pairing of end segments; more proximal pairing is non-homologous; buckles in central position. (D) complete pairing with double cross. (E) Derivatives of double cross with decreasing homologous pairing in one or both of the terminal segments. (After BURNHAM et al., 1972)

ends associate, but then it is a different combination of two chromosomes than in the first case. Pairing only at the ends leads to large unpaired interstitial segments, which may pair non-homologously at pachytene, forming a buckle when the segments are different in length. Pairing of homologous interstitial segments with unpaired ends results in non-homologously paired distal segments and, when the segments are unequal in length, to single terminal stretches of chromosome (Fig. 3.5). The relative frequencies of double crosses (or their immediate derivatives in the form of simpler quadrivalents) and the two types of bivalent can be estimated and related to the location of the break points in the chromosome arms. Since chromosomes 1 and 5 differ in length, and usually unequal segments were exchanged, the two members of the pairs differed in length sufficiently to make them recognizable. In Table 3.5 a summary of the results of BURNHAM et al., (1972) is given. Five groups of intercrosses were distinguished, depending on the length of the interstitial segments. A translocation with both interstitial segments smaller than 30% of the arm was designated by 0; with one smaller than 30%, the other longer by 1; both longer than 30% by 2. In the Table the total number of interstitial segments larger than 30% is given for each intercross. When all four are short, this number is 0; when all are long, it is 4: there are five classes. Only one cell was found with bivalents which had their interstitial segments paired. This occurred in the class with four long interstitial segments. In all other cells analyzed either bivalents with the end segments paired or a multivalent were observed. This demonstrates that pairing initiation is concentrated distally, and that centromere pairing initiation is excluded. It is interesting to note that the quadrivalent frequency increases as the length of the interstitial segments increases. One particular interchange with the break points far out in the arms (1L.92 and 5L.82), which in a particular combination produced the exceptional bivalent just mentioned, was given special attention. Three intercrosses with this interchange LL-5 were studied in respect

Table 3.5. Pachytene pairing in combinations of two reciprocal translocations at a time between chromosomes 1 and 5 of maize, in which opposite arms are involved, cf. Figs. 3.5, 3.6 (BURNHAM et al., 1972). Grouped according to the number of interstitial segments larger than 30% of the length of the arm concerned. Each of the two translocations in each combination may have 0, 1 or 2 of such interstitial segments. Total of 24 translocations in 36 combinations

Combination of interstitial segments >30%	Cells	Cells with "bivalents" paired at		% with "bivalents"	% with quadrivalents
		end	diff. segm.		
0 × 0	16	16	0	100	0
0 × 1	289	260	0	87.2	12.8
1 × 1 } 0 × 2 }	281	100	0	37.1	62.9
1 × 2	197	55	0	30.8	69.2
2 × 2	453	17	1	5.6	94.4
Total	1,236	448	1		

to the behavior of the multivalent (Fig. 3.5). With complete homologous pairing, a double cross is formed: both parental interchanges contributed one cross. If one of the two interchanged segments of one translocation fails to pair while the other segment does, a T-shaped connection is formed. It can be combined with a cross when the other translocation still has both interchanged segments paired, or with another T when that translocation also has only one interchanged segment paired. The fourth type of quadrivalent arises when one translocation has failed to pair in both interchanged segments, while the other still forms a cross. BURNHAM *et al.* (1972) give the frequencies of these configurations for intercrosses between LL-5 and translocations with smaller interstitial segments (SS types). Especially with SS-4 (1S.66 and 5S.29) but also with SS-5 (1S.69 and 5S.71), the 2-cross and 1-cross configurations were scarce. This suggests that pairing initiation is infrequent in the short terminal segments but much higher in the adjacent segments. By dividing the chromosome arms in 10 units of equal length and by trial and error fitting of various probabilities of pairing to the data, the probabilities of Fig. 3.6A were obtained for the average arm of chromosomes 1 and 5. Then, for instance, for LL-5 with breaks at 1L.92 and 5L.82, the probability of initial pairing in 1L is 0.3 and in 5L it is 0.8

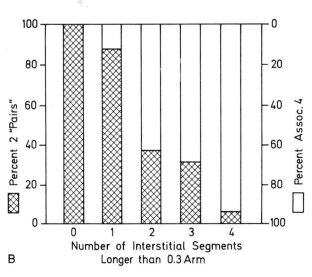

Fig. 3.6. (A) Approximate average probabilities of pairing initiation in segments equal to 0.1 the length of either arm of chromosomes 1 and 5 of *Zea mays*. (B) The relation between association in pairs (or in quadrivalents) and the length of the interstitial segments in the combination of two reciprocal translocations between the same chromosomes in *Zea mays* (cf. Fig. 3.5). There are four interstitial segments; in the abscissa the number of those of the four that are longer than 0.3 of the arm involved, is given. (BURNHAM et al., 1972)

(i.e. $0.5 + 0.3$). In 113 cells of the LL-5 × SS-5 combination the observed frequencies and those expected on this basis were:

Initial pairing in LL-5	Observed	Expected frequency		Expected number
Both interchanged segments	25	0.3×0.8	$= 0.24$	27
Interchanged segm. of 1 ⎫	69	$0.3 \times (1-0.8) = 0.06$ ⎫	0.62	70
Interchanged segm. of 5 ⎭		$(1-0.3) \times 0.8 = 0.56$ ⎭		
In neither	19	0.7×0.2	$= 0.14$	16
Total	113		1.00	113

The frequency of quadrivalents is an indication of the occurrence of "secondary" sites of pairing initiation. In a number of combinations these sites are in interstitial segments, as otherwise only bivalents can be formed (Fig. 3.5) It appears that, as already stated above, the further removed the breakpoints are from the centromere, the higher the multivalent frequency. In Fig. 3.6 B this relation is shown graphically. In general, the frequency of pachytene associations of four increased with increasing length of the interstitial segments, but a number of interactions must operate, since the relation is not always straightforward. Therefore, although Fig. 3.6 A, giving the approximate probabilities of pairing initiation for the different segments of maize chromosomes, has general significance, the relations in any actual situation are more complex.

Temperature extremes may markedly alter chromosome pairing, as do some genetic factors. There are also other modifiers. In maize, a variant of chromosome 10 (K 10) exists, which has a pronounced effect on chromosome pairing and on subsequent crossing-over. RHOADES and DEMPSEY (1966) and RHOADES (1968) reported on increases in crossing-over in the presence of K 10 in several stocks of maize, and especially the paracentric inversion In 3 b heterozygotes were interesting. In the proximal region recombination between the genes Gl and Lg was increased four times compared to the heterozygote without K 10. Pollen abortion, resulting from exchange in the inversion segments (bridge and fragment formation) was increased by 70%. At pachytene, pairing near the inversion was much more intimate (Fig. 3.7). In In 3a a threefold increase in crossing-over between Gl and Lg was noted, accompanied by a similar increase in the number of anaphase I cells with a fragment but without bridge. Homozygotes

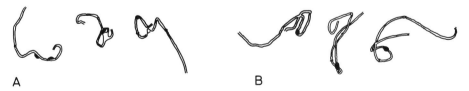

A   B

Fig. 3.7 A and B. Pachytene pairing in an inversion heterozygote (chromosome 3) in *Zea mays*. (A) Non-homologous and incomplete pairing in normal stocks. (B) More complete (homologous) pairing in stocks carrying abnormal chromosome 10 (K 10). Drawn after RHOADES, 1968

for the inversion, and even normal plants showed an increase in crossing-over in this region, explained by more effective pairing in a region which normally pairs well according to visual standards, although this may not be the effective type of pairing required for exchange. The fact that heterochromatin is relatively abundant in proximal regions may be significant.

Another indirect indication for the effect of K 10 on pairing resulting in an increase in crossing-over was an increase in ring quadrivalents and a decrease in trivalents with univalents in plants heterozygous for translocation T 6–9 b, combined with an increased recombination between *Sh* and *Wx* in chromosome 9:

	% rings	% chains	% trivalents	% recombination *Sh–Wx*
Normal 10	19.4	48.4	32.2	11.5
K 10 heterozygote	68.8	30.4	0.8	24.0

3.1.4 Indirect Observation: Post-Pachytene Configurations

When MAGUIRE'S (1965, 1966, 1967) contention is accepted that pairing is invariably followed by crossing-over, the analysis of post-pachytene configurations is as relevant for pairing as the analysis of pachytene. The natural exceptions would be (a) the cases of significant movement of chiasmata when diakinesis and metaphase stages are studied, (b) the cases of genetically or otherwise conditioned desynapsis where a preliminary committment to exchange is not realized and (c) the cases of non-homologous pairing, either in the absence or, probably less frequently, in the presence of proper homologues. In all other cases, pairing would commit to exchange. For a number of reasons, however, (see also RHOADES, 1968) this seems to go too far. Probably it is better to state that even in perfect diploids homologous pairing is far from complete and presents a restriction on crossing-over. This results in a strong correlation between the occurrence of pairing in any segment, and the occurrence of exchange in that segment. When pairing has the opportunity to proceed from the point of initiation, more than one chiasma can be formed in the paired segment. On the other hand, not every paired segment will necessarily contain a chiasma, even when the partners are homologous. The dispute has not been settled satisfactorily, and for the time being it still seems reasonable to assume simply that the probability of a chiasma being formed is correlated with the extent of pairing and that with an increase in length of the paired segment, the probability of more than one chiasma being formed increases. It is certain that when there is a chiasma, there must have been pairing, but the frequency of chiasmata is determined by other factors (for instance interference) to such an extent that it may not *a priori* be stated that the number of chiasmata in a segment is simply proportional to the extent of pairing in that segment. And absence of chiasmata is no proof of absence of pairing. The more closely MAGUIRE'S (1965, 1966, 1967, 1972) suggestions that chiasmata necessarily follow contact between chromosomes approach reality, the more relevant chiasma studies are for pairing, but a general correlation is probably all one can hope for.

The number of reports on chiasma observations that can be used in the analysis of pairing, is limited, and refers to a limited number of approaches. *Bivalent interlocking* (DARLINGTON, 1965) can be of considerable interest. A serious drawback is its low frequency even after experimental induction (BUSS and HENDERSON, 1971). When also dependent on chiasma formation, a meaningful *quantitative* analysis of pairing is hardly possible.

There are two situations in which, because of a very low frequency of homologous pairing and formation of chiasmata in a relatively large proportion of the paired sites, the study of frequency and localization of chiasmata can be used in an analysis of pairing. This is especially so when preliminary observations have given an impression of the degree of the correlation between chiasma frequency and extent of pairing. Nevertheless, any variation in chiasma formation as such decreases the value of the interpretation. These situations are: (a) Haploids and interspecific hybrids with reasonable frequencies of chiasmata in duplicated segments. When the level of non-homologous pairing is known to be low, the correlation between pairing and crossing-over can be estimated and, (b) true asynaptics, where complete homology is present, but for some (genetic) reason pairing is only very incompletely realized.

In the "coral" haploid of *Nicotiana tabacum*, an analysis of chiasma distributions (Table 3.2) indicated that there was one segment which formed chiasmata especially frequently. It was assumed to be the known "coral" duplication. In LEVAN'S (1942) rye haploids, pairing at pachytene was usually almost complete and mainly non-homologous. Whereas the maximum chiasma frequency may be an indication of the extent of homology, variation in chiasma frequency and distribution may reflect variation in pairing, although it may also be variation in chiasma formation as such. Two haploid plants showed a difference in chiasma frequency: in one there were 0.6869 chiasmata per cell, in the other only 0.0928. In the latter, the distribution over cells was random (Poisson), in the first, it appeared that the class of one chiasma per cell was greatly in excess (Table 3.6). LEVAN (1942) reasoned that the variation in chiasma formation may here have two components: (a) pairing in one chromosome segment which had a great probability of having a chiasma, and (b) pairing in the rest of the genome. When for (a) a chiasma frequency of 0.4 was taken, and consequently for (b) 0.2869, random distribution of pairing and of chiasmata in the remainder of the genome gave a good fit with the observed frequencies. This means that about 60% (=0.4 out of 0.6869) of the chiasmata may be formed in a special region of high pairing frequency, the rest being formed at random. A similar approach applied for various haploids in earlier reports made it possible to find the conditions under which these too would give a good fit. It is probable that in such cases relatively large duplications exist, which in some genotypes have the capacity to pair and form chiasmata, whereas in other genotypes they fail to do so. Possibly in cases of low chiasma frequencies pairing initiates predominantly in non-homologous, for instance heterochromatic segments, progressing from there on, and this decreases the chance for homologous regions to find each other.

In true (partial) asynaptics, analysis of metaphase chiasma frequencies can lead to comparable conclusions. SWAMINATHAN and MURTY (1959) reported on normal, and both partial and almost completely asynaptic plants segregating among derivatives of a *Nicotiana rustica* × *N. tabacum* cross ($2n = 48$) (Table 3.7). As in the haploids, plants with very low chiasma frequencies (almost complete asynapsis) had bivalent distributions (over cells) conforming to a Poisson series. With higher chiasma frequencies deviations occurred which suggested that some bivalents paired more frequently than others, and perhaps these were specific

Table 3.6. Distribution of chiasmata in two haploids of *Secale cereale*, rye; one with high, the other with low chiasma frequency. The second fits a Poisson distribution. If in the first it is assumed that one locus pairs with a frequency $a = 0.40$, the rest pairing at random, a good fit is found. If there is no locus with a high pairing frequency ($a = 0.00$), the fit is very poor (LEVAN, 1942)

Average chiasmata per cell	Total cells		Number of chiasmata per cell						χ^2
			0	1	2	3	4	5	
0.6859	1,252	Cells observed	557	552	125	14	4	—	
		exp. $a = 0.40$	563.9	537.6	131.0	17.7	1.8	—	0.8603
		exp. $a = 0.00$	629.9	432.7	148.6	34.1	5.9	0.8	
0.0928	1,304	Cells observed	1,189	110	4	1	—	—	
		exp. $a = 0.00$	1,188.5	110.3	5.1	0.1	—	—	0.0087

bivalents. Since tobacco is an allotetraploid, there is a possibility that the chromosomes of one genome are more liable not to pair than those of the other. If this difference between genomes would be a critical factor, the number of bivalents would be expected to be close to 12 ($=x$). The prominent peak at 5 bivalents per cell suggests that only a few (but more than one) chromosomes had a high frequency of pairing. Although quantitatively not analyzable, pachytene pairing reflected the metaphase association levels: highly asynaptic plants had almost no pairing, partly asynaptics had a much better, although far from complete pachytene pairing.

Effect of external influences on pairing may also be studied at metaphase, when quantitative analysis at pachytene is not possible. In order to distinguish between effects on pairing and effects on chiasma formation, at least some qualitative check at pachytene is desirable. Even then, part of the effect can be on chiasma formation, unless it can be proven that the influence has been exerted exclusively in a period at which exchange cannot yet have been initiated. Even when MAGUIRE'S (1965, 1966, 1967, 1972) assumption is accepted, that pairing commits to exchange, there must be a period of attraction preceding effective contact. Temperature treatments and chemicals have been applied to early or even premeiotic stages in order to test their effect on the meiotic processes. LEVAN (1939), for instance applied colchicine to inflorescences of *Allium cernuum* ($2n = 14$) and collected metaphase I cells after intervals of one to several days. After a relatively short interval, pachytene pairing was severely reduced, and after a 5 days' interval this was reflected in a very low chiasma frequency at metaphase I:

Number of bivalents: 0 1 2 3 4 5 6 7 Total
Number of cells: 8 7 5 4 5 2 0 0 31.

The normal number is seven bivalents per cell. This decrease is considered a reflection of disturbed pairing, perhaps by inhibiting the formation of microtubuli functioning in the process of primary attraction.

Table 3.7A and B. Distribution of chiasmata in partial asynaptic and asynaptic Nicotiana derivatives (SWAMINATHAN and MURTY, 1959)

(A) Chiasmata at diakinesis: average number per cell of configurations with 0, 1, 2 or 3 chiasmata

Plant	No. of cells	Chiasmata per bivalent				Av. chiasm. per cell	Type
		0	1	2	3		
28/1, 2	45	0.27	1.18	19.17	3.55	50.17	normal
22/1; 30/1	30	0.61	1.68	19.33	2.32	47.60	normal
3/3	20	37.20	4.65	0.75	0	6.15	partial asynaptic
7/2	15	14.16	11.69	5.23	0	22.15	partial asynaptic
5/2	25	46.32	0.84	0	0	0.84	asynaptic
6/9	26	46.92	0.39	0.15	0	0.69	asynaptic

(B) Numbers of cells with 0, 1, 2 etc. bivalents

Plant		Bivalents per cell											χ^2
	0	1	2	3	4	5	6	7	8	9	10	11 12	
5/2	29	17	5	5	—	—	—	—	—	—	—	— —	
Poisson	26.45	19.84	7.44	1.86									0.66
6/9	32	4	1	0	1	—	—	—	—	—	—	— —	
Poisson	29.30	7.62		1.08[a]									2.75
3/3	0	2	2	1	8	29	6	5	—	—	—	— —	
Poisson		5.67[a]	8.75	11.22	10.80	8.32	5.34	5.34					69.13[b]
7/2	0	2	0	0	0	0	1	8	8	2	15	0 5	
Poisson		5.86[a]				5.25[a]	5.22	5.6	5.35	8.19		6.32[a]	16.60[b]

[a] Classes combined.
[b] Signif. at 5% level.

Similar and more extensive analyses, also including temperature treatments, have been carried out in experiments on pairing and chiasma formation in wheat, usually without analysis of pachytene. They have been an important contribution to the understanding of the pairing relations between homologous and homoeologous chromosomes and genetic factors determining chromosome behavior in allopolyploids (BAYLISS and RILEY, 1972a, b; DOVER and RILEY, 1973; DRISCOLL and DARVEY, 1970; DRISCOLL et al., 1967).

3.2 Competitive Situations: Localization of Pairing Initiation Points; Pairing Differentiation

3.2.1 Trisomics

3.2.1.1 Primary Trisomics

Pairing in competitive situations is essentially different from pairing in non-competitive situations. Although long distance attraction can involve more than two chromosomes at a time and even short stretches of synaptonemal complex have been observed to form between more than two homologues (MOENS, 1968b; COMINGS and OKADA, 1971), the rule is that final pairing is between only two homologues at any location on the chromosome. This implies that a choice has to be made between potential partners when more than two are available. Of course, pairing has to be effected, and therefore variation in pairing intensity or capacity plays a role here as well as in non-competitive situations. Competition, however, introduces an entirely new dimension.

Three to some extent interdependent parameters can be distinguished in competitive pairing: the potential and realized number of points of pairing initiation and their location; partner exchange; partner preference.

Pachytene pairing is in principle analyzable in favorable subjects. In EINSET'S (1943) example, which has been considered before, a pachytene trivalent frequency of 0.036 was observed for chromosome 3. Such data are very rare and often not very reliable. Usually information on diakinesis or metaphase association is all that is available. Dependence on chiasma formation then is a serious limiting factor, as small paired segments may go undetected when they fail to have a chiasma. Trivalent pairing, for instance, may go undetected when a bivalent with an univalent appear at diakinesis. Yet EINSET'S (1943) data on diakinesis-metaphase indicate a trivalent frequency of 0.786 for chromosome 3; the average for eight trisomes was 0.857.

Primary trisomy marks the chromosome involved and at the same time permits analysis in the most simple example of competitive pairing between homologous chromosomes. It has been considered in Section 2.3.1 and for the different pairing configurations the reader is referred to Table 2.35 and Fig. 2.29. The frequency of trivalent pairing was represented by f in Eq. (2.33) in which the metaphase trivalent frequency is given as $f \cdot a \cdot b$ and Eq. (2.34) in which the frequency of ring bivalents is $(1-f) \cdot a \cdot b$. From these f can be derived. For estimating the frequency of ring bivalents derived from the trisomic set of chromosomes, information on the non-trisomic chromosomes is required. This introduces a considerable sampling error. It was also shown that with trivalent pairing the frequency of chiasmata is lower than with bivalent pairing, and this reduces the estimate for f. Finally, with trivalent pairing, the point of partner exchange is variable and when it is close to the end, an interstitial chiasma may be formed without a distal chiasma. Then a ring bivalent is formed, again leading to an underestimation of f. In general, therefore, f will be estimated too low. In the example given for rye, the value of 0.623 probably indicates that f is close or equal to $\frac{2}{3}$, which is expected when pairing starts at both ends of the chromosomes and only infrequently in between, and when it is random between the three chromosomes. Thus, in this example no significant preferential pairing between specific partners can be demonstrated.

If f rises over $\frac{2}{3}$, the cause may either be a number of pairing initiation points larger than two (Fig. 2.29) or a preference for pairing between two specific chromosomes at one end and the other two at the other end. When f is smaller than $\frac{2}{3}$, either pairing initiation concentrates on one chromosome end, or there is preferential pairing between two specific chromosomes in both ends and perhaps also in other segments. Both systems may vary genetically. In a homozygote or highly inbred strain, pairing preferences will be absent and the value of f is solely determined by the localization system of pairing. Only then can this system be analyzed. When $f=0$, there is a single point of pairing initiation (Fig. 2.29). When $f=\frac{2}{3}$, there are two per chromosome, separated by enough distance to permit free partner exchange. With $\frac{2}{3}>f>0$, partner exchange is limited, and the points of pairing initiation must be close together or some are relatively ineffective. When $f>\frac{2}{3}$, there must be more, their number and/or distance increasing with increasing f.

3.2.1.2 Telocentric and Secondary Trisomics

Telocentric trisomics have been fully considered in Section 2.3.3. Examples of estimates of f have been given, and the limitations of telocentric trisomics for studying chromosome pairing have been discussed.

Secondary trisomics (isochromosome trisomics) are somewhat less common. Four homologous arms are available for pairing, two of which are attached

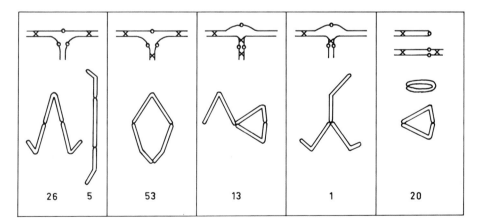

Fig. 3.8. Pairing configurations (trivalent, and bivalent + univalent) in a secondary trisomic, and the metaphase I configurations formed with the chiasma indicated in the pairing configurations. The numbers refer to the frequencies of the different configurations observed by BELLING and BLAKESLEE (1924) in a secondary trisomic of *Datura stramonium* (118 cells)

to the same centromere (Fig. 3.8). Again the largest possible configuration is a trivalent, and unlike in the other trisomics, it may be in the form of a ring. With equal probability of association of all four homologous arms, twice as many trivalents will be formed as the combination of a bivalent with a self-paired univalent. The point of partner exchange may be at the centromere or anywhere in the arm. With full pairing, the configuration is symmetric, but since pairing around the point of partner exchange tends to lapse, a somewhat asymmetric arrangement is possible. For detection at diakinesis-metaphase chiasmata are required in the critical segments. The isochromosome can sometimes be recognized at meiosis: when the normal chromosome has markedly different arms, the isochromosome will be either larger or smaller than the normal chromosome, and its centromere is in the middle. If the isochromosome resembles the normal chromosomes, the combination of a rod univalent with a rod bivalent may either be due to iso-normal or to normal-normal pairing. However, this dilemma arises only with (sub-)metacentric chromosomes, which infrequently lack chiasmata in both arms. Whenever a ring bivalent or a ring univalent appear, the bivalent must be composed of two normal chromosomes. If not directly recognizable, this class can be estimated by the subtraction method. The number of non-classifiable configurations tends to be relatively small.

As with all trisomic configurations, trivalent pairing with partner exchange near the end, when chiasmata in the terminal segments are absent, will not be recognized in the secondary trisomic, and therefore the frequency of trivalent pairing tends to be underestimated. The extent to which this occurs depends on the pattern of localization of partner exchange and of chiasmata, and cannot easily be determined. Since, in general, terminal pairing initiation is common, and often chiasmata are terminally concentrated, this class of configuration will usually not be disturbingly large. There is not much information on the meiotic behavior of secondary trisomics. As an example the oldest reported case (BELLING and BLAKESLEE, 1924) is given in Fig. 3.8 (Datura). All categories observed could be classified. The number of (ring) univalents for the isochromosome is small (about one in six), thus apparently the connection of the two arms at the centromere does not help the arms to become paired preferentially. Pairing, therefore, must have begun near the ends and some more locations along the chromosome. Whether preferential pairing or a large number of initiation sites are responsible for this high trivalent association, cannot be determined. Rather variable behavior in several combinations of an isochromosome with telocentrics in rye have been described by SYBENGA and VERHAAR (1975).

3.2.1.3 Telocentric Trisomy in Telocentric Substitution

The combination of telocentric trisomy with telocentric substitution for metacentric chromosomes is especially favorable for analyzing pairing preferences. One of the two homologues is substituted by the corresponding telos, the other is normal. In addition, the telo for one of the two arms is present as an extra chromosome: for this arm, two telocentric copies and one associated with a normal second arm are available for pairing (Fig. 3.9 A). With complete pairing, besides cases with partner exchange, where all three arms are associated, there are: (a) the association of the two telos, giving a small bivalent together with a heteromorphic bivalent formed by the association of the normal chromosome with the telo for the other arm, (b) the association of either one of the telos for the trisomic arm with the normal chromosome which has its second arm associated with the corresponding telo. This results in a trivalent. One telocentric chromosome remains univalent. When all three homologous arms have the same probability of finding a partner, the ratio between (a) and (b) will be 1:2. Any deviation from this ratio must be the result of differences between the chromosomes in respect to their attraction system.

There are two major complications: partner exchange and low chiasma frequency. With chiasmata in all paired segments, partner exchange leads to branched configurations (Fig. 3.9 B) in which all three are associated, and then it is difficult to determine if there is any preference for association between specific partners at specific sites. When distal pairing dominates, as is common, partner exchange will be infrequent, and enough unbranched configurations remain. If partner exchange is not infrequent but as a result of chiasma localization or interference seldom leads to branched configurations, uncertainties may arise. In Fig. 3.9 B the three modes of pairing with partner exchange are shown. There are always two paired segments. If a chiasma is formed in only one, in mode 1. a trivalent with a univalent will result (two possibilities). With modes 2. and 3., either a trivalent with univalent, or two bivalents will be formed. With random chiasma formation, the 1:2 ratio is maintained (two bivalent sets *vs.* four trivalents), and it is not difficult to see that even localization of chiasmata will not introduce a deviation.

Low chiasma frequencies result in high univalent frequencies (Figs. 3.9 A, B) and the impossibility to distinguish between different pairing types. As long as chiasma failure is random, the only consequence is that larger numbers of cells have to be scored. When, however, special pairing combinations lead to loss of chiasmata, a bias may be introduced.

In an experiment in rye (SYBENGA, 1972 b, c), the genotype of the normal chromosome (Table 3.8) was varied systematically (three genotypes). The genotypes of the telocentric chromosomes, although from the same background, were variable due to recombination. The first plant was a sister plant of one of the ancestors of the others. In those others the normal chromosome was derived (by crossing) from a specific inbred line and thus constant and known. The possibility exists that the pairing initiation sites (zygomeres, SYBENGA, 1966 b; cf. Fig. 3.12) in the normal and telocentric chromosomes were exchanged by recombination, i.e. the pairing specificity (or activity) present in the normal chromosome may have been transferred to the telos tested in later generations. Then the zygomere types of the telos in the later generation plants (Table 3.8) correspond with those of the normal chromosome in the early generation plant.

As appears from Table 3.8, the branched types and univalents were relatively infrequent. The short arm trisomic will be considered first: in the plant related to the ancestor of the other two, the frequency of two bivalents greatly exceeded

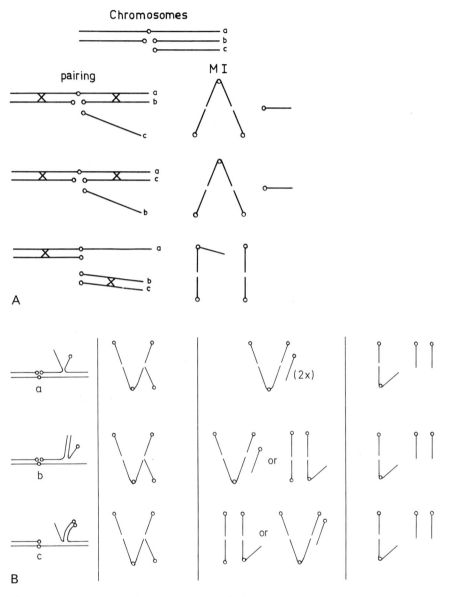

Fig. 3.9 A and B. Telocentric trisomy combined with telocentric substitution. (A) The chromosomes, and the pairing and M I configurations. With equal probability of pairing between any two of the three (a, b, c) homologous arms, twice as many trivalents (with univalents) as bivalent pairs are expected. (B) The three possible modes of pairing with partner exchange in the trisomic arm. In all three, formation of chiasmata in both segments leads to a branched trivalent. When only one chiasma is formed, a will result in a trivalent with univalent; b results in a trivalent with univalent when the chiasma is formed in the proximal segment, and in a bivalent pair when it is formed in the distal segment; c results in the opposite. Thus, independent of chiasma localization, the 2:1 ratio between trivalents (with univalent) and bivalent pairs is maintained. Absence of chiasmata in the trisomic arm results in two telocentric univalents. (SYBENGA, 1972c)

Table 3.8. Metaphase I configuration frequencies in rye plants with telocentric substitution of one of the satellite chromosomes combined with telocentric trisomy. Normal chromosome from unknown source and from inbred lines 001 and 015. For calculation of the trivalent/bivalent ratio the infrequent configurations D and F were combined with C and E respectively. All ratios for the short arm telo-trisomics significantly different from 2, the others only when pooled. (SYBENGA, 1972 c). Cf. Fig. 3.9

Plant	Short arm telo trisomics							Long arm telo trisomics		
	66300-175	69492-1	69495-29	69496-2	69496-10	69496-66	69496-69	69496-12	69496-26 a	b
Cells	1.000	600	500	1.000	1.000	600	800	800	500	500
Normal chrom.	?	001	001	015	015	015	015	015	015	015
Configurations										
A	0.028	0.028	0.004	0.035	0.040	0.025	0.066	0.085	0.100	0.070
B	0.001	0	0	0.001	0.002	0	0.004	0.003	0.002	0.002
C	0.469	0.568	0.638	0.699	0.646	0.637	0.648	0.565	0.568	0.566
D	0	0.003	0.002	0.001	0.011	0.005	0.010	0.006	0.008	0.024
E	0.366	0.208	0.136	0.103	0.105	0.098	0.069	0.330	0.314	0.326
F	0.005	0.003	0.002	0.002	0.002	0	0	0.005	0.006	0.010
G	0.131	0.187	0.216	1.58	0.189	0.233	0.201	0.009	0.002	0.002
H	0	0.002	0.002	0.001	0.005	0.002	0.004	0	0	0
C+D/E−F	1.264	2.706	4.638	6.667	6.140	6.525	9.536	1.704	1.800	1.756
Bound arms trisomic	0.869	0.812	0.782	0.841	0.806	0.765	0.795	0.986	0.984	0.964
Disomic	0.994	0.992	0.994	0.995	0.980	0.993	0.983	0.991	0.998	0.998

the expected $\frac{1}{3}$, suggesting that either the telos had a special preference for each other, or that the normal chromosome was inefficient (compared to the telos) in finding a partner, by reduced pairing activity or any other reason. In the plants in which the normal chromosome was derived from either one of the two inbred lines, the reverse was found. This can mean either that a special affinity existed between the normal arm and one of the telos, or that the normal arm was especially active in pairing. It is possible that the relatively inefficient zygomeres of the normal arm of the ancestral type had been transferred to one (or even both) of the telos. It seems that especially inbred line 015 had a relatively efficiently pairing normal chromosome, with trivalent/bivalent ratios of 6.667; 6.140; 6.525 and even 9.536. When this same chromosome was made telo-trisomic for the *other* arm, the ratio was well below 2, showing a reduced pairing efficiency (Table 3.8). This demonstrates that in respect to finding a partner, the two arms are entirely independent.

Several more combinations of telocentric and normal chromosomes can be made. More complex situations can be created when normal chromosomes are combined with both telocentrics and isochromosomes. More specific associations can then be recognized (SYBENGA and VERHAAR, 1975).

3.2.1.4 Translocation Trisomics and Telocentric Translocation Trisomics

The combination of a structural rearrangement with trisomy may make it possible to partition a chromosome into specific segments and to study pairing initiation and pairing preferences in these segments separately. Simultaneously, the effect of the rearrangement on pairing can be studied. It may not always be possible to separate the two. A simple combination of a chromosomal rearrangement with trisomy can be constructed when a reciprocal translocation is combined with a complete extra chromosome. This is the case with tertiary trisomy, where to a normal karyotype a translocation chromosome is added, or a normal chromosome to a translocation homozygote. Another type is translocation trisomy, where one chromosome of the translocation is added to the heterozygote. There are four types which in principle are equivalent (Fig. 3.10). An example is given by SYBENGA (1967b), corresponding with Fig. 3.10A. Arms a and b are trisomic, and when only the end segments are considered, there are three possible combinations for a ($a_1 - a_2$; $a_1 - a_3$; $a_2 - a_3$) and also three for b. Combined these give nine different pairing combinations. In one of these, the two structurally homologous chromosomes are paired at both ends, forming a bivalent. The rest of the complex is a trivalent. When the rearrangement impedes pairing between rearranged and normal chromosomes, favoring pairing between structurally homologous chromosomes, this particular configuration must occur in more than $\frac{1}{9}$ of the cases. This test may seem simple, but is greatly complicated by the fact that the frequencies of the different pairing configurations can be determined at diakinesis-metaphase only when chiasmata occur in all terminal segments capable of pairing. This is not necessarily realized. When, for instance, the bivalent of the two structurally homologous chromosomes fails to have chiasmata in one arm, a rod bivalent results. A rod bivalent is formed also

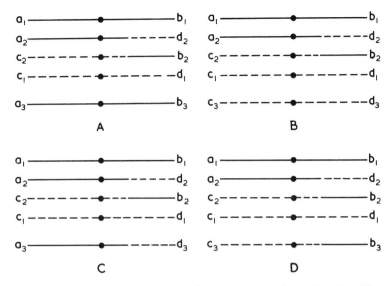

Fig. 3.10A–D. The four possible translocation trisomics of a reciprocal translocation. The extra chromosome is shown at the bottom of each set

when in a pairing configuration of five, chiasmata are absent in the arm pair connecting the third and second (or third and fourth) chromosomes. It is necessary, then, to have information on chiasma frequencies from some other source, for instance the normal heterozygote, before conclusions on the pairing behavior can be based on the configuration frequencies in translocation trisomics. When genetic and environmental conditions are not entirely the same, or when other sources of error are present, this information is of limited value. In addition, extra chromosomes tend to reduce chiasmata in critical segments, mainly as a result of partner exchange. When the chromosomes are very different in shape, it may be possible to recognize specific associations, but this is only infrequently the case at diakinesis-metaphase. It appears, therefore, that small effects on pairing cannot be traced, and only strong preferential pairing is expressed in significant deviations from randomness. In the example given, a complicated analysis was required to show that preferential pairing, if at all present, would have to be assumed to be quite weak for the translocation studied. For the same reason, distribution of, and variation in pairing initiation points could not be analyzed.

Much more effective are telocentric translocation trisomics (Fig. 3.11). The two advantages are that the extra chromosome is readily recognized at meiosis and that it can be made to correspond only with an arm in which a break has occurred. A disadvantage is that there is not one single critical configuration which is increased in frequency as a result of induced preferential pairing. In the example given here (SYBENGA, 1972b), translocation 248 of rye (with a break in the satellite, close to the end) was combined with a telocentric for the arm with this break, the short arm of the satellite chromosome. When all end segments capable of pairing are associated, three configurations can

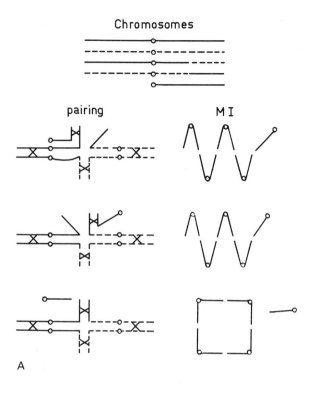

A

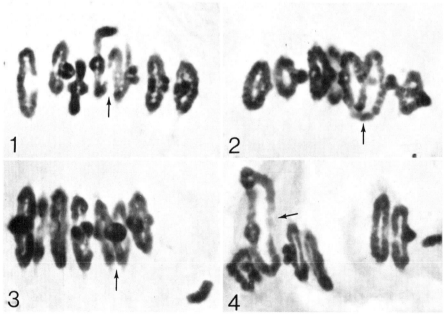

B

be formed: two are a chain of five (one with the telo attached to the normal arm, the other with the telo attached to the translocation arm), and one is a ring of four with a telocentric univalent (Fig. 3.11). With preferential pairing between the normal arm of the complete chromosome and the normal telocentric arm, the first type of chain of five increases in frequency at the expense of the other chain and of the ring + univalent. Again, chiasma frequencies lower than 1 disturb the analysis, while in addition partner exchange, resulting in branched multivalents, may be a complication. In the latter case, either the telo has not been able to make a choice and has paired with both, or it has paired at the interstitial segment and not at the end. In view of the small interchanged segment, the latter is the most probable in this example. As regards the absence of chiasmata, its frequency is of consequence only for the single arm involved both in trisomy and in the translocation. The number of different critical configurations is relatively small, and all that has to be scored is whether the telo is attached terminally, or interstitially, or not associated at all. Information from other sources in respect to this translocation indicated that with pairing between two normal arms (one of the telo, the other of the normal chromosome) a maximum of 98% are bound by chiasmata, with an average of 0.95. It can occasionally be lower. An association between a normal and a translocation arm leads to chiasmata in a maximum of 0.8 of the cases and an average of 0.6 approximately (again occasionally much lower). With random pairing (Fig. 3.11) a maximum terminal metaphase association of $\frac{1}{3} \times 0.98 + \frac{1}{3} \times 0.8 = 0.59$ (average $\frac{1}{3} \times 0.95 + \frac{1}{3} \times 0.6 = 0.52$) is expected. When this is compared with the observed values (Table 3.9) it is seen that in only one case is the average expected value realized. In all others, terminal association is much higher, in one case (0.946) almost as high as the frequency of being bound for the entire, non-translocated arm. This shows that in this particular plant practically complete preferential pairing between the two structurally homologous arms has occurred. The logical implication is that this translocation *as such* induces considerable preferential pairing. The segregation observed in spite of the presence of a constant genotype for the normal complement (Table 3.9), results from segregation of zygomere systems in the telocentric and/or the translocation chromosomes. A few more conclusions present themselves: if pairing initiation were restricted to the terminal segments, no preferential pairing could have been introduced by the translocation. Some zygomeres must be present in the interstitial segment, which can also be concluded from interstitial chiasmata in this and in the

---

Fig. 3.11 A and B. Telocentric translocation trisomic. (A) The chromosomes, the three pairing configurations and the corresponding M I configurations when chiasmata have been formed in all paired segments. The interstitial segments are assumed not to pair with the telos unless pairing is initiated in the end segment and can proceed from there. Thus the paired segment may be longer when the telo pairs with the normal arm than when it pairs with the translocated segment. Partner exchange in the telocentric arm combined with formation of chiasmata on both sides of the point of partner exchange leads to interstitial attachment of the telo (cf. B. 2). (B) Photomicrographs of telocentric translocation trisomics in *Secale cereale*. *1*: Chain of five, telo terminally attached. *2*: Ring with telo interstitially attached. *3*: Ring of four (alternate orientation) with univalent telo. *4*: Ring of four (adjacent orientation) with univalent telo. Multivalents indicated (arrows)

Table 3.9. Attachment of the telocentric chromosome in translocation telocentric trisomics of rye, in which the normal chromosome is derived from different inbred lines (001 and 015), and in which the pairing characteristics of the other chromosomes segregate. Translocation 248. Cf. Fig. 3.11 (SYBENGA, 1972 b)

Number of cells	1,000	500	500	500	500
Telo attachment:					
terminal	0.827	0.758	0.800	0.946	0.520
interstitial	0.043	0.074	0.110	0.012	0.058
free	0.130	0.168	0.090	0.042	0.422
Origin normal chromosome	001	015	015	015	015
Bivalent arms bound	0.964	0.953	0.975	0.986	0.962
(average other chromosomes)					

Expected maximum terminal association with random pairing $\frac{1}{3} \times 0.98 + \frac{1}{3} \times 0.8 = 0.59$.
Expected average $\frac{1}{3} \times 0.95 + \frac{1}{3} \times 0.6 = 0.52$.

disomic translocation heterozygote. When wide variation in preferential pairing occurs, this cannot be due to variation in interstitial zygomeres in the translocation chromosome, since this segment is practically immune to recombination. Since

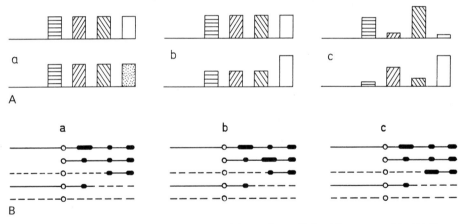

Fig. 3.12 A and B. Variation in zygomere activity and specificity (diagrammatic). (A) Four points of pairing initiation (zygomeres) interdependently acting in the regulation of long-distance attraction. (a) In two homologous chromosomes by mutation one of the four has differentiated into a new "allele". Attraction is still effective between the other three, but when in a tetrasomic both types are present twice, there will be preferential pairing between like types. (b) The specificity has not altered, but the activity has. (c) Almost complete differentiation as a result of alterations in activity in all four zygomeres. In a tetrasomic, complete preferential pairing results. (B) Variation in zygomere activity in the three homologues of a telocentric translocation trisomic. The normal chromosome has the same constitution in all three cases; it has relatively active zygomeres in the interstitial and terminal segments. (a) Telocentric and translocation chromosomes identical in zygomere make-up. There is some preferential pairing between the two non-translocation arms. (b) An especially active zygomere in a more distal part of the telocentric increases preferential pairing. (c) Transfer of this zygomere (by crossing-over) to the translocated segment reverses the effect. (SYBENGA, 1972 b)

the normal chromosome is kept constant (derived from inbred lines), the variation must be confined to the telo. All telos used were derived from a single translocation trisomic parent, and therefore, recombination between the telo and the normal chromosome in *this* parent must have been the source of the variation. For recombination in the terminal segment both normal and translocation chromosome in the parent were available. It is suggested (Fig. 3.12 B) that in the normal chromosome a rather strong zygomere is located in the interstitial segment, which is not present in the translocation and telocentric chromosome, where a weaker allele takes its place (Fig. 3.12 Ba). This difference is enough to cause a considerable basic preferential pairing. A specially active zygomere in the telocentric increases preferential pairing (Fig. 3.12 Bb). When this occurs in the translocated segment, but not in the telocentric, preferential pairing is counteracted (Fig. 3.12 Bc), as in the plant with random pairing (Table 3.9). A similar pattern can be constructed, although less readily, when specificities rather than intensities are considered the cause of zygomere variation. A variation in activity is the more probable explanation, and can result in patterns entirely comparable to variation in specificity (SYBENGA, 1973 b and Fig. 3.12 A c).

3.2.1.5 Inversion Trisomics

Inversion trisomics are in some respects comparable to translocation trisomics. When pachytene analysis is unreliable (as is usually the case with complicated configurations) crossing-over between segments paired must be relied upon. For inversions, anaphase bridges are usually scored more conveniently than metaphase chiasmata. Pericentric inversions do not form bridges, but crossing-over in (paired) inversion segments can occasionally be scored at anaphase as chromatid exchange. A detailed analysis of pairing in a paracentric inversion trisome of maize was performed by DOYLE (1963), who used In 3a in the long arm of chromosome 3. The breakpoints, according to RHOADES and DEMPSEY (1953), were at 0.4 and 0.95 from the centromere (pachytene length). The locus of the gene A was placed at 46 cross-over units from the proximal break point and at 15 units from the distal break point. The gene Sh_2 is closely linked to A, (0.25 units). Three approaches for detecting possible preferential pairing between the structurally identical chromosomes were tried: (1) modification of gene segregation; (2) modification of diakinesis trivalent frequency and (3) modification of anaphase bridge frequency.

The first approach is of limited interest in the present context, although some interesting deviations from segregations expected with random pairing were observed, confirming the results of the analysis of the meiotic configurations.

The trivalent frequency in 427 cells from seven InNN trisomes was reduced from 68.9% in the controls to 56.6% ($\chi^2 = 15.9$; P < 0.0005). The control frequency was close to the theoretical $\frac{2}{3}$ for pairing initiation at two independent locations. How far this reflects the actual pairing situation cannot be traced since no pachytene analysis was reported. Nor was the type of bivalent described in the non-trivalent associations. If this bivalent were predominantly a rod bivalent, chances are that several of the bivalents are a result of breakdown of trivalents as a result of chiasma failure. This would affect both homologous and heterologous pairing. In fact, the non-inversion short arm of chromosome 3 is only 25 map units long, which is equivalent to appr. $\frac{1}{2}$ chiasma: in almost half the cases this arm must have been free at diakinesis, which is even more than corresponds with 69% trivalents. Thus there is an apparent increase in crossing-over in the short arm, which may be a result of reduced across-centromere interference.

Anaphase bridge frequencies were a better criterion. In a duplex heterozygote (InNN) with random pairing, $\frac{2}{3}$ of the associations of the inversion segment may be expected to be heterosynaptic (In-N), and $\frac{1}{3}$ homosynaptic (N-N). In heterosynaptic association a chiasma leads to an AI bridge or an AII bridge (when an additional, disparate, chiasma is formed in the interstitial segment). Two complementary chiasmata give a double bridge. With random pairing in the trisome, $\frac{2}{3}$ of the bridge frequency is expected compared to that in the disome, where heterosynaptic pairing is the only mode possible. The following bridge frequencies were reported:

	Anaphase I			Anaphase II	
	No bridge	Single bridge	Double bridge	No bridge	Bridge
Diploid (InN)	667	394	18	1940	126
	(61.8%)	(36.5%)	(1.7%)	(93.9%)	(6.1%)
Trisome (InNN)	935	156	4	1365	37
	(85.4%)	(14.2%)	(0.4%)	(97.4%)	(2.6%)
Expected ($\frac{2}{3}$ × diploid)		24.3%	1.1%		4.1%

Since the frequency of the two types of double cross-overs is not reduced more than that of single cross-overs, the chiasma frequency is probably not seriously affected by trisomy, and the lower bridge frequencies must be almost entirely due to reduced heterosynapsis. The total frequency of cells with any crossing-over in the inversion is $0.365 + 0.017 + 0.061 = 0.443$ in the diploid. The expected value for the trisomic is $\frac{2}{3} \times 0.443 = 0.295$. The observed frequency is $0.142 + 0.004 + 0.026 = 0.172$. It is reasonable to assume that heterosynapsis has been reduced by a factor $\frac{0.172}{0.295} = 0.583$. The trivalent frequency was reduced less. If the non-inversion (short) arm of chromosome 3 had no reason to pair preferentially, with homosynapsis of the normal homologues of the inversion arm still in only $\frac{1}{3}$ of the cells the short arms of these particular chromosomes would pair. In $\frac{2}{3}$ the third short arm would pair with either one of the other two. Thus, in spite of preferential pairing of the inversion arm, the expected trivalent frequency would still be $\frac{2}{3}$. Only when the arms are interdependent, i.e. when preferential pairing would extend into the non-inversion arm, would a reduction in trivalent pairing result. The alternative is that inversion heterozygosity reduces crossing-over sufficiently to cause a number of arms not to be associated at diakinesis. The reduction of trivalent frequency would then have no relation to preferential pairing.

3.2.2 Autopolyploids

3.2.2.1 Autopolyploids and Autopolysomics without Structural Rearrangements

Autotriploids are equivalent to multiple trisomes and will not be treated separately. Autotetraploids have been studied very extensively and of the numerous reports only a few will be considered. The principle of pairing is similar to that of trisomics (Fig. 2.25) with the difference that there is a pairing partner for every chromosome. In respect to pairing initiation, again three systems may be distinguished:

1. Pairing initiates at one site only: two bivalents result from each set of four homologues.
2. Pairing initiates at two independent sites (for instance one at either end) and a ratio of two quadrivalents to one set of two bivalents is expected (Table 3.10). When the two points are interdependent, more bivalents may result.
3. Pairing initiates at more than two points, and depending on their number and interaction, more than $\frac{2}{3}$ quadrivalents may be formed. On the basis of the number of quadrivalents, therefore, a rough estimate of the number of sites of pairing initiation can be made. Complications are: (a) Any preferential pairing between specific chromosomes will reduce the frequency of quadrivalents. In the extreme case only bivalents are formed, and the polyploid in fact is an allopolyploid. When the preferences are different for different segments of the same chromosome, the multivalent frequency may be increased. (b) Low number or pronounced localization of chiasmata leads to breakdown of pairing multivalents into bivalents. In autotetraploid *Allium porrum*, LEVAN

Table 3.10. Pairing configurations in an autotetraploid (tetrasomic). Both arms of each chromosome are present four times. When pairing starts predominantly at the chromosome ends, each arm has three possibilities: a_1 pairs with a_2 (then a_3 pairs with a_4); a_1 pairs with a_3 (a_2 with a_4); a_1 pairs with a_4 (a_2 with a_3). The same is true for b. When partner exchange is free between the two points where pairing starts in each chromosome, there is a total of 9 combinations, all with the same probability of occurring. Of these, three give two bivalents (II + II) and six will give one quadrivalent (IV). When pairing initiates also in interstitial segments, more quadrivalents can be formed. (SYBENGA, 1972a)

Chromosomes:		a_1———O———b_1	
		a_2———O———b_2	
		a_3———O———b_3	
		a_4———O———b_4	
	b_1-b_2	b_1-b_3	b_1-b_4
	b_3-b_4	b_2-b_4	b_2-b_3
a_1-a_2 a_3-a_4	II + II	IV	IV
a_1-a_3 a_2-a_4	IV	II + II	IV
a_1-a_4 a_2-a_3	IV	IV	II + II

(1940) reported a pachytene quadrivalent frequency of 1–8 per cell with one, occasionally two partner exchanges randomly located in the quadrivalents. At diplotene they were still recognizable, but at diakinesis, all quadrivalents fell apart into pairs of bivalents. Yet there were almost always two chiasmata in each bivalent. These, however, were localized in short segments on both sides of the centromeres. If partner exchange occasionally took place in the centromeric region, it would interfere with chiasma formation, and only one would be formed. Of the 10 quadrivalents observed at metaphase in 380 cells, some were nevertheless due to partner exchange between two chiasmata in the centromeric region, others to occasional non-localized chiasmata. In related autotetraploids of Allium where chiasmata are more randomly located, about 100 times as many quadrivalents are present at metaphase. Also in autotetraploid teosinte (SHAVER, 1962) localized chiasma formation is the main cause of bivalent association at diakinesis. In most autopolyploids chiasma formation tends to be reduced, partly because of genetic unbalance, partly because of partner exchange. This results in a lower number of multivalents at diakinesis-metaphase than corresponds with the pairing pattern: not all paired segments have chiasmata. (c) Interaction between individual "zygomeres" may cause neighboring zygomeres to attract the same partner. They act like blocks rather than individually. (d) In result similar but causally different is the mechanical limitation in frequency of partner exchange. Even though there may be many points of pairing initiation each attracting different partners, the stiffness of the chromosomes prevents all the theoretically possible changes of partner from being realized.

A predominant diakinesis-metaphase quadrivalent frequency of about $\frac{2}{3}$ was reported by MORRISON and RAJHATHY (1960a, b) for a number of autotetraploid plants. A lower frequency (around 0.48) was observed by McCOLLUM (1958) and numerous other authors in Dactylis glomerata, and higher frequencies by VENKATESWARLU (1950) and GILLES and RANDOLPH (1951) in maize. At pachytene, VENKATESWARLU never found more than two points of partner exchange, i.e. not more than three initiation sites were independently active (SVED, 1966). Even higher were the frequencies cited by JOHN and HENDERSON (1962) for a number of insects with high chiasma frequencies. In the latter cases it is clear that more than two points of pairing initiation must have been active independently in the same pairing period in the same chromosome, but the actual number cannot be estimated on the basis of quadrivalent frequencies alone. There are more than the number of attachment sites on the nuclear membrane. The number of potential initiation sites can be determined with even less success.

When the original number of quadrivalents per cell cannot be determined at metaphase because of failure of chiasma formation, there are ways to derive this number from the relative frequencies of the different configurations present. A few simplifications are inevitable but even though the outcome of such a simplified analysis may not in all respects be correct, with careful interpretation it can give a reasonable impression of pairing behavior, chiasma frequencies and their interrelations.

Considering only organisms with pronounced distal chiasma localization (which lack chiasmata in interstitial segments), the following configuration types can be distinguished (cf. Figs. 2.9; 3.13): ring quadrivalents (frequency rq); chain quadrivalents (frequency cq); trivalents with univalent (frequency t); ring bivalents (frequency r); open bivalents (frequency o) and univalent pairs (frequency u). The first three must necessarily be derived from quadrivalent pairing; the ring bivalents are assumed to be derived from bivalent pairing, and the open bivalents and univalent pairs can be formed both after quadrivalent and bivalent pairing. With metacentric chromosomes these configurations are maintained into MI; with acrocentric chromosomes diakinesis is the latest stage which still contains all information. When chiasmata occur also in interstitial segments, the classification of the multivalents runs parallel with that discussed for the reciprocal translocations (cf. 2.2.4.2.1), but the possibility arises that a number of ring bivalents are formed after quadrivalent pairing.

The frequency of quadrivalent pairing is f, in analogy with the analysis of the trisomes. The probabilities of chiasmate association of the end segments A and B are named a and b respectively. The frequencies of the different configurations can be given in terms of a, b and f. Since in quadrivalent pairing the length of the end segment varies between almost 0 and the length of an entire arm, while in addition partner exchange may reduce the chiasma frequencies, different frequencies of chiasmate association must be assumed for the quadrivalent (a' and b') and bivalent pairing situations (a and b). The averages can then be named $\bar{a}$ and $\bar{b}$.

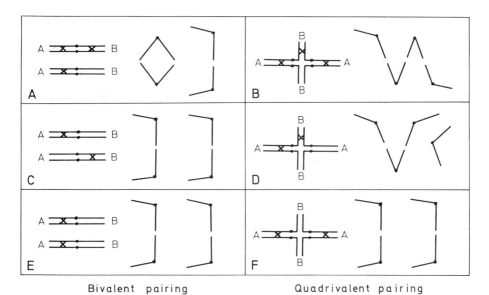

Fig. 3.13A–F. Pairing and M I configurations in a tetrasomic. (A), (C), (E) Bivalent pairing and examples of the resulting combinations of bivalents and/or univalents, depending on the frequency and location of chiasmata. (B), (D), (F) Quadrivalent pairing and examples of the resulting types of quadrivalents, trivalents with univalent, bivalents and univalents, depending on the frequency and location of chiasmata

Since frequencies of configurations comprising different numbers of chromosomes are scored (quadrivalents *vs.* bivalents) the expressions refer to the total numbers of chromosomes involved rather than the numbers of configurations. The frequency of chromosomes involved in ring quadrivalents then is:

$$rq = f a'^2 b'^2. \tag{3.1}$$

The derivation of the frequency of the chain quadrivalents is somewhat more complicated: $f\{2(1-a')a'b'^2 + 2a'^2b'(1-b')\}$ or,

$$cq = 2f(a'b'^2 + a'^2b' - 2a'^2b'^2). \tag{3.2}$$

The trivalent + univalent frequency is $f\{4a'b'(1-a')(1-b')\}$ or,

$$t = 4f(a'b' - a'b'^2 - a'^2b' + a'^2b'^2). \tag{3.3}$$

The ring bivalent frequency is

$$r = (1-f)\,ab. \tag{3.4}$$

The derivation of the frequency of open bivalents is considerably more complicated. They may originate from both quadrivalent and bivalent pairing. In the first case, there may be two of them:
$a'^2(1-b')^2 + b'^2(1-a')^2$, or one bivalent and two univalents:
$a'(1-a')(1-b')^2 + b'(1-b')(1-a')^2$. Both multiplied by f. In the second case:
$a^2(1-b)^2 + b^2(1-a)^2 + 2ab(1-a)(1-b) + ab(1-a)b + ab(1-b)a + b(1-a)^2(1-b) + a(1-a)(1-b)^2$, multiplied by $(1-f)$.
The total frequency of open bivalents then is:

$$o = f(a'+b'-4a'b'+a'^2b'+a'b'^2) + (1-f)(a+b-2ab). \tag{3.5}$$

Finally, along similar lines, the frequency of univalent pairs can be derived as

$$u = f(1-a'-b'+a'^2b'+a'b'^2-a'^2b'^2) + (1-f)(1-a-b+ab). \tag{3.6}$$

The sum $rq+cq+t+r+o+u$, of course, equals 1: there are six equations and five degrees of freedom. By substitutions between Eqs. (3.1), (3.2) and (3.3) it appears that

$$f a'b' = \frac{t + 2(cq + 2rq)}{4},$$

and since $f a'^2 b'^2 = rq$, it is easy to derive

$$f = \frac{(t + 2cq + 4rq)^2}{16rq}. \tag{3.7}$$

Further,

$$a'b' = \frac{4rq}{t+2cq+4rq}$$

and

$$a'b' = \frac{2cq+8rq}{t+2cq+4rq}.$$

In 50 cells of *Tradescantia virginiana*, $4n=24$ (SYBENGA, 1975) 65 ring quadrivalents (260 chromosomes); 92 chain quadrivalents (368 chromosomes); 9 trivalents with univalent (36 chromosomes); 161 ring bivalents (322 chromosomes); 101 open bivalents (202 chromosomes) and 6 univalent pairs (12 chromosomes) were observed. This corresponds with $rq=0.217$; $cq=0.307$; $t=0.030$; $r=0.268$; $o=0.168$; $u=0.010$. According to (3.7) f was estimated to be 0.658, close to the "random" value $0.667 (=\frac{2}{3})$. The number of multivalents observed was only 0.554. The average values for chiasmate association of the two arms were $\bar{a}=0.946$ and $\bar{b}=0.683$ (cf. 2.1.2.2). Solving a' and b' from Eqs. (3.1) and (3.2) gave $a'=0.950$ and $b'=0.605$. Substituting these values in Eqs. (3.4) and (3.5) led to equal roots for a and $b=0.885$. This is an improbable result since a' is approximately equal to $\bar{a}$. Both a and b were therefore derived anew from the relations $f\times a' +(1-f)\times a=\bar{a}$ and $f\times b'+(1-f)\times b=\bar{b}$. This resulted in $a=0.938$ and $b=0.830$ using the values of f, a', b', $\bar{a}$ and $\bar{b}$ as given above. Substitution in Eqs. (3.1)–(3.6) give configuration frequencies almost exactly equal to those observed demonstrating that these estimates were quite acceptable.

The complications encountered when estimating a and b probably result from the low frequency of the univalent class, on which the difference between a and b largely depends. This is not the only complication in such analyses. The example given above consisted of two fixations of the same clone. One was used earlier for demonstrating how $\bar{a}$ and $\bar{b}$ could be estimated (2.1.2.2). This fixation had an excess of ring bivalents and a shortage of ring quadrivalents which made it impossible to estimate chiasmate association after bivalent pairing. This effect was apparently compensated by the second fixation. Such irregularities are of great genetic and developmental interest and deserve further study.

The fact that the value of f is rather higher than the frequency of quadrivalents observed shows that numerous quadrivalents must have fallen apart because of lack of chiasmata. Or, with higher chiasma frequencies there would have been more quadrivalents. McCOLLUM (1958) indeed found a correlation between the average number of chiasmata and the average number of quadrivalents in *Dactylis glomerata*, both between plants and between cells within an anther. The correlations were very significant but not high and the regression coefficients were far below 1. When grouping cells with low chiasma frequencies separately from cells with high frequencies, the correlation in the first (0.2976) were slightly but very significantly higher than in the second (0.2811), as expected. The relatively low correlations indicate that other factors besides chiasma frequency play a role.

Since high quadrivalent frequencies have been correlated with irregular meiotic behavior in autotetraploids (resulting in partial sterility), it has been suggested that selection for low chiasma frequencies would improve fertility. Since this, however, results in an increase in the frequency of univalents, which are even more deleterious for fertility, this would not be a sound procedure. It would be better to select for the proper behavior of multivalents, or for restriction of pairing initiation to a single block, or for chiasma localization. Selection for higher chiasma frequencies in autotetraploid rye led to an increase in quadrivalent frequencies. According to TIMMIS and REES (1971) there still was an excess of bivalents, even of ring bivalents, which they ascribed to more than random bivalent pairing. In spite of the fact that these authors purposely selected cells with high chiasma frequencies in order to be certain that no quadrivalent breakdown could be the cause of this bivalent excess, their approach must be viewed with caution. Since quadrivalents in general have a lower chiasma frequency than bivalents, the selected group automatically excludes cells with high frequencies of (potential) quadrivalents.

There in no *a priori* reason for all chromosome sets in an autopolyploid to have the same pairing behavior. JOHN and HENDERSON (1962) found an exclusive bivalent type of pairing for the short chromosome of Chorthippus, whereas the large chromosomes had high multivalent frequencies. When the different homologous sets can not be distinguished, a statistical approach must be relied upon. McCOLLUM (1958) compared the distribution of quadrivalent numbers over cells, in Dactylis tetraploids, with a binominal distribution and found excellent correspondence in most of the 37 plants studied. In two, there was a deviation, but this is expected on the basis of chance alone. Two examples, one with a comparatively high and one with a comparatively low quadrivalent frequency (both showing a good fit with a random distribution) are given in Table 3.11. SHAH (1964) found the same good correspondence for trivalent distributions in triploid Dactylis. The suggestion (based on the relatively low multivalent frequencies) that the tetraploid species *Dactylis glomerata* might be a segmental allotetraploid, is clearly not justifiable. It is an autotetraploid with relatively low quadrivalent pairing. In hybrids low quadrivalent association

Table 3.11. The distribution of quadrivalents over cells at Metaphase I compared with a binomial distribution in two *Dactylis glomerata* autotetraploids, one an artificial tetraploid of subspecies *lusitanica*, the other an artificial tetraploid of the hybrid between subspecies *smithii* and *lusitanica*. (From McCOLLUM, 1958). Average quadrivalent frequency expressed per 4 homologues

Plants		Numbers of quadrivalents per cell								Number of cells	Average quadriv. frequency	Variance
		0	1	2	3	4	5	6	7			
1585-9	Observed	1	3	11	46	49	40	15	5	170	0.575	1.7106
	Expected	0.3	4.1	16.3	36.9	50.0	40.5	18.3	3.6	170.0		1.6444
sm × 1638-4	Observed	41	18	6	0					65	0.066	0.4315
	Expected	40.3	19.9	4.2	0.6					65.0		0.4399

may have its cause in preferential pairing, but there is no reason why this should be restricted to specific chromosomes.

In autotetraploids (in which all chromosomes are tetrasomic) the different chromosomes can often not be distinguished and the pooled data may not permit a satisfactory analysis. In tetrasomics for single chromosomes, there is no doubt in respect to the chromosome studied. When in addition the two arms can be distinguished, rather detailed information may be obtained.

It would seem that the example given in 2.1.2.2 of B-chromosome tetrasomics in rye (SYBENGA and DE VRIES, 1972) should be especially favorable for analysis. The method used for Tradescantia virginiana, however, is not applicable, since multivalent formation based on pairing and chiasma formation in two arms of the same chromosome is not sufficiently frequent here: the short arm only rarely has a chiasma. Multivalent formation depends primarily on partner exchange in the long arm and on simultaneous distal and interstitial chiasma formation. Since this combination is scarce, partly caused by interference across the point of partner exchange, the frequency of multivalents at diakinesis-metaphase is low. In this respect it resembles short acrocentric chromosomes with a negligible short arm, as described by JOHN and HENDERSON (1962). Estimates of the number of pairing initiation sites, their interdependence and relative activity and specificity depend on the availability of partner exchange. Since partner exchange is revealed only by the virtue of rare simultaneous occurrence of interstitial and terminal chiasmata, which are interdependent, little can be done.

A special form of tetrasomy is presented by the combination of two isochromosomes (attached chromosomes; symmetric compound chromosomes) without further homologous chromosomes (Fig. 3.14; SYBENGA and VERHAAR, 1975). The pairing may be as a bivalent or as two univalents. In the first case there is the possibility of "partner exchange" and the formation of interstitial chiasmata, but this is not necessary. In fact all transitions between "normal" bivalent pairing and (double) univalent pairing are possible. This is a complication always encountered with autotetrasomics, resulting in a somewhat underestimated quadrivalent pairing frequency, since with a very distal point of partner exchange chiasmata will infrequently be formed in the distal segment. An extreme example of this effect was reported by LEVAN (1940) for Allium porrum which is an autotetraploid with pronounced proximal chiasma localization. Quadrivalents are observed at pachytene but fall apart at later stages, and at M I practically only bivalents remain. In the example of two isochromosomes in rye, with predominantly distally localized chiasmata, the effect is probably small. The isochromosomes were derived from the short arm of the satellite chromosome. The other arm of this chromosome was present as a pair of telocentrics, and no normal homologue was available. If it may be assumed that the majority of the ring univalents have resulted from true within-univalent pairing, the following simplified notation (cf. Fig. 3.14) may be introduced: bivalent pairing f; univalent pairing $(1-f)$; frequency of being bound with bivalent pairing a; with univalent pairing a'. Now the frequency of two ring univalents can be given as $(1-f) \cdot a^2$. One ring and a rod univalent have a frequency $2(1-f) \cdot a \cdot (1-a)$ and two rod univalents $(1-f)(1-a)^2 + f(1-a')^2$. The frequency of ring bivalents (in addition to the

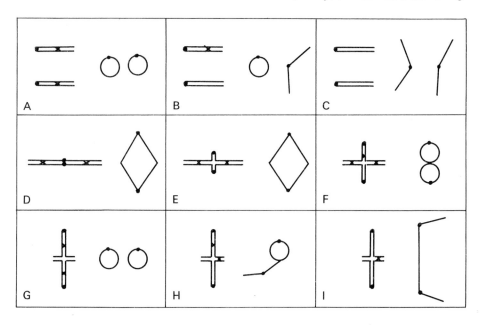

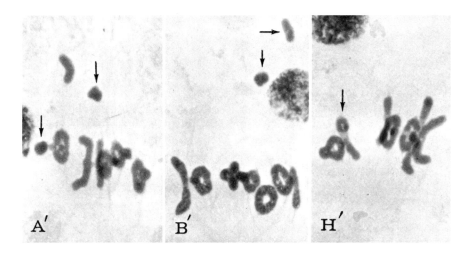

Fig. 3.14A–I. Pairing and M I configurations of the combination of two isochromosomes. (A), (B), (C) Univalent pairing and different combinations of ring and rod univalents at M I, depending on the presence or absence of chiasmata in the self-paired univalent. (D–I) Bivalent pairing with shifting point of partner exchange and examples of the resulting types of M I configuration. (A^1), (B^1), (H^1) The M I configurations of (A), (B) and (H) respectively, as observed in *Secale cereale* (SYBENGA and VERHAAR, 1975). The isochromosomes are derived from the short arm of the satellite chromosome. There is no normal satellite chromosome. The long arm is represented by two telocentrics, which form a small bivalent. In (A^1) two ring univalents (arrows). The rod univalent is another chromosome, the homologue of which is not in the picture. (B^1) A ring and a rod univalent of the isochromosomes; seven bivalents of the others. (H^1) A "frying-pan" bivalent of the isochromosomes; seven bivalents of the others (one not in the figure)

infrequent "figure eight" resulting from an interstitial chiasma, Fig. 3.14F) is $f \cdot a'^2$; that of open bivalents (with the frying pan) is $f \cdot 2a' \cdot (1-a')$. These are four independent equations and three unknowns. In fact there are six different association types (Table 3.12, Fig. 3.14), each of which may have its own chiasma frequency, but only the two mentioned will be considered. The one degree of freedom left after estimating a, a' and f is not sufficient to estimate the chiasma frequencies for all of these association types, and it seems best to combine the two univalent pairing types for estimating an average univalent chiasma frequency, a, and to combine the bivalent pairing types for estimating a'. The one degree of freedom left is not used. The analysis is demonstrated on two plants of rye, using several fixations for each. The frequency of two ring univalents in plant 71451-20 (3,100 cells) was 0.103, that of one ring univalent with an open univalent 0.357. Thus $(1-f) \cdot a^2 = 0.103$ and $2(1-f) \cdot a \cdot (1-a) = 0.357$.

Table 3.12. Frequencies of Metaphase I configurations of the two isochromosomes in two plants of rye in which the satellite chromosomes were replaced by two telocentrics for the long arm and two isochromosomes for the short arm. For configuration types see Fig. 3.14. (From SYBENGA and VERHAAR, 1975)

Plant	71451-20	71452-12	Arms bound
Configuration:			
A	0.103	0.436	2
B	0.357	0.290	1
C	0.375	0.116	0
D	0.028	0.033	2
E	0.096	0.083	1
F	0.012	0.029	2
G	0.029	0.013	1
Number of cells	3,100	1,150	
Average number of bound arms a	0.384	0.691	
a^2	0.147	0.477	
Observed config. with both arms bound	0.143	0.498	

From this $a = 0.366$ and $f = 0.231$. The same procedure can be followed for the bivalents: $f \cdot a'^2 = 0.040$ and $2f \cdot a' \cdot (1-a') = 0.125$. This leads to $a' = 0.390$ and $f = 0.263$. There are slight differences between the univalent and bivalent pairing results, but these are not disturbing. An average frequency of being bound for the tetrasomic arm can be derived as $a = 0.384$. The total number of configurations with both arms bound is $0.103 + 0.028 + 0.012 = 0.143$; this corresponds well with $a^2 = 0.147$, indicating absence of interference across the centromere.

The results for plant 71452-12 (1,150 cells) were rather different. The average a was much higher: 0.691; $a^2 = 0.477$ and the observed frequency of configurations with both arms bound was 0.498, not too deviating, at most suggesting some negative interference. The further analysis was less regular. From $(1-f) \cdot a^2 = 0.436$

and $2 \cdot (1-f) \cdot a \cdot (1-a) = 0.290$, a was derived as 0.750 and f as 0.225. For the bivalents, $f \cdot a'^2 = 0.062$ and $2f \cdot a' \cdot (1-a') = 0.096$ leads to $a' = 0.564$ and $f = 0.195$. The differences both between the values for a and for f are considerable. Since the class of open univalents (no chiasmata) is derived from both types of pairing, it is worth while to give it some more attention. Its frequency is $0.116 = (1-f) \cdot (1-a)^2 + f(1-a')^2$. When for a and a' the values as derived above are substituted, f is solved as 0.420. This value is excessively high and can only be explained when a' as used in the formula is too high and a too low. This means that the number of univalents without chiasmata derived from bivalent pairing is lower than expected on the basis of the ratio between ring and open bivalents, or that there are not enough ring bivalents. Similarly, there must be too many pairs of ring univalents in relation to the number of one ring-one rod univalent combinations. This is understandable when in this particular plant, in contrast with the other, the higher frequency of chiasmata observed are primarily interstitial chiasmata, perhaps formed after more pronounced proximal pairing initiation. Then, subsequent or even simultaneous distal pairing initiation would lead to rather distally located partner-exchange, interfering with chiasma-formation in this region. Interference would then reinforce this pattern, and instead of ring bivalents, pairs of ring univalents arise. The values estimated for f might thus be much too low, even that estimated from the chiasmate univalents. Yet all estimates for f remain well below the expected $\frac{2}{3}$, and this may partly be due to the two arms always being in close proximity. This, however, is in contrast with what was observed in secondary trisomic pairing. The value of a is also much lower than expected. One cause may be a reduction because of partner exchange. Another, initial proximal pairing combined with distal chiasma localization.

3.2.2.2 Autopolyploids with Inversions

In his analysis of preferential meiotic pairing in maize, DOYLE (1963) compared tetraploid simplex and duplex paracentric inversion heterozygotes. In the first case, preferential pairing is excluded, since the inversion chromosome, if it pairs at all, must always pair with a normal chromosome. Bridge frequency is representative for crossing-over in the inversion under tetraploid conditions, and can be used as a control value for the frequency with which bridges are formed in a duplex heterozygote following heterosynapsis. The expected bridge frequency in the duplex in absence of preferential pairing then is $\frac{2}{3}$ (frequency of heterosynapsis) $\times 2$ (there are two such heterologously synapsed pairs) $\times$ the simplex bridge frequency. In DOYLE'S (1963) example the frequency (Bs) of cells with a bridge in the simplex was 0.337. This times $\frac{2}{3} \times 2$ equals 0.449. The duplex bridge frequency observed (Bd) was only 0.154. The frequency of heterosynapsis (h), equal to $\frac{2}{3}$ with random pairing, was estimated as follows: $2 \times h \times Bs = Bd$, or $2 \times 0.337\, h = 0.154$ and $h = 0.228$. This is considerably lower than the expected 0.667. The segregation ratios were altered accordingly, but since preferential pairing primarily involves one set of arms, not to the same extent. In fact, as in the trisome, only when preferential pairing affects the chromosome as a whole (i.e. both arms in a metacentric) is it expected to affect segregation. When at least part of the chromosome associates at random and always has a chiasma, there is no preferential segregation. If the non-inversion part associates at random but in a significant proportion lacks chiasmata and there by fails to contribute in determining segregation, a corresponding proportion of the preferential pairing resulting from the rearrangement will be expressed as a tendency towards disomic segregation. In DOYLE'S (1963) example the simplex testcross gave an excess of recessives: the ratio was 1:1.2. In the duplex AAaa × aaaa testcross without the inversion, in 4,413 gametes 20.83% recessives (aaaa) appeared, or a 3.82:1 ratio, again an excess. In the same heterozygote, but now duplex for the inversion (AA in the inversion), in 3,038 gametes 10.80% recessives were found or a ratio of 8.26:1. When aa was in the inversion there were 12.56% recessives or a 6.74:1 ratio. The difference with the "normal segregation" is highly significant and suggests rather strong preferential pairing. There are a few complications, one being that double reduction is lowered (it tends to lead to more recessives), but quantitatively this is of minor importance. Of more consequence is loss of the dominant allele as a result of deficiency for the acentric fragment in cases where deficient gametes are tolerated. This is expected to be relatively frequent in the female tetraploid heterozygote, but less so in the male, where pollen competition eliminates deficient gametes. In the reciprocal cross, the deviation in fact was only slightly smaller: aaaa ♀ × AAaa ♂ with AA in the inverted segment gave 13.40% recessives or a 6.44:1 ratio in 4,295 gametes. With aa in the inverted segment in 9,406 gametes 14.03% recessives, or a 6.13:1 segregation was observed. The corresponding control without the inversion gave a 4.05:1 segregation. The difference remains considerable, and it is clear that not only for the arm, but also for the entire chromosome, preferential pairing has been induced by the inversion. A rough comparison between the degree of preferential pairing as estimated from anaphase bridge frequencies and from preferential segregation is possible. The segregation ratios must first be corrected for deviations not due to preferential pairing. One is loss of the dominant allele as a result of the formation of acentric fragments containing this allele. The total frequency of AI fragments (cf. Table 3.13) equals $0.118 + 2 \times 0.007 + 2 \times 0.015 + 2 \times 0.050 = 0.262$. In the tetraploids with AA in the inversion chromosomes (Fig. 3.15) the dominant allele is lost in $\frac{1}{4}$ times this 0.262 and when AA is in the normal chromosomes in $\frac{3}{4} \times 0.262$ of the cells, since region II is three times the genetic length of I. Since the deficient genotype is phenotypically identical to the recessive type, cross-overs in region II or non-cross-overs lead to the normal 25% recessive gametes, but those in region I to 37.5%, the difference being 12.5%. Thus, with AA in the inversion chromosomes there is an excess of gametes with the recessive allele, amounting to $\frac{1}{4} \times 0.262 \times 12.5\% = 0.82\%$. With AA in the normal chromosomes the excess of recessives is $\frac{3}{4} \times 0.262 \times 12.5\% = 2.45\%$. This refers to the female as the heterozygote, which permits the recovery of deficient gametes in tetraploids.

Another disturbing factor is double reduction, also increasing the frequency of gametes with only recessive alleles. Although the locus of A is rather far out in the chromosome, double reduction is not expected to be quantitatively of much importance since it depends on the frequency of multivalents with adjacent orientation for the arm involved. This is probably infrequent, for instance because the other arm of chromosome 3 is short and lacks chiasmata in a large number of cases.

Table 3.13 (A). Anaphase I and II configurations of simplex and duplex tetraploid inversion heterozygotes of maize (DOYLE, 1963)

| | No bridge | Anaphase I | | | Anaphase II | |
		Single bridge	Double bridge	Two bridges	No bridge	Bridge
Simplex	656	324	9		885	63
(In/N/N/N)	(66.3%)	(32.8%)	(0.9%)		(93.4%)	(6.6%)
Duplex	754	103	6	13	460	24
(In/In/N/N)	(86.1%)	(11.8%)	(0.7%)	(1.5%)	(95.0%)	(5.0%)

Table 3.13 (B). Gene segregation in tetraploid inversion heterozygotes and controls (DOYLE, 1963). Genes marked with * are in an inversion chromosome

Cross	No. of gametes tested	% a	Ratio A:a
Aaaa × aaaa	983	54.63	1:1.20
A*aaa × aaaa	922	54.92	1:1.12
aaaa × Aaaa	2,950	51.12	1:1.05
aaaa × A*aaa	1,885	51.62	1:1.07
AAaa × aaaa	4,413	20.83	3.82:1
A*A*aa × aaaa	3,038	10.80	8.26:1
AAa*a* × aaaa	6,360	12.56	6.74:1

DOYLE (1963) did not extend the analysis beyong this point, but a few more details can be worked out. Thus, with random pairing, two thirds heterologous pairing is expected, and the frequency of recessive gametes $R = \frac{1}{4}$. With preferential pairing, the proportion of heterologous pairing (h) decreases, and so does R: $R = \frac{1}{4} \times \frac{3}{2} \times h$. The value for R must be corrected for loss of dominant alleles and for deviation from normal 3:1 segregation of the control (3.82:1); this latter correction is $\frac{4.82}{4}$, and therefore the corrected $R' = \frac{4.82}{4} \times (R + 0.0082)$ for the segregation with the heterozygote as the female and the dominant alleles in the inversion chromosomes: $R' = \frac{4.82}{4} \times (0.1080 + 0.0082) = 0.140 = \frac{1}{4} \times \frac{3}{2} \times h$, or $h = 0.373$. This is much higher than h estimated from the anaphase bridge frequency (0.228) but still considerably lower than the random $\frac{2}{3}$. When estimating h from the situation with A in the normal chromosome, an even higher value ($h = 0.483$) is obtained. The difference is considerable, suggesting that the correction for loss of the A allele is of dubious value. This loss should be greater with *A* in the inversion chromosome (higher percentage of recessives expected) than with *A* in the normal chromosome (Fig. 3.15), but the opposite is observed (Table 3.13). Factors such as preferential segregation of non-crossover chromatids, reduced recovery of deficiency gametes, deficiencies caused by breakage of the bridge rather than by loss of the fragment play an important but unpredictable role. Nevertheless it is clear that there is preferential segregation and this may well be based on preferential pairing. However, preferential pairing in a single set of arms in an autotetraploid is not expected to result in preferential segregation when all arms capable of pairing do so and form a chiasma. The second arm of chromosome 3 is short and does not have a chiasma in a large proportion of the cells, and this is a basis for some preferential segregation. It must then be accompanied by an increased bivalent formation for this particular chromosome which in itself is merely a reflection of reduced chiasma formation in the short arm, and not of preferential pairing. It is important to note that preferential pairing in inversion heterozygotes with metacentric chromosomes cannot be studied

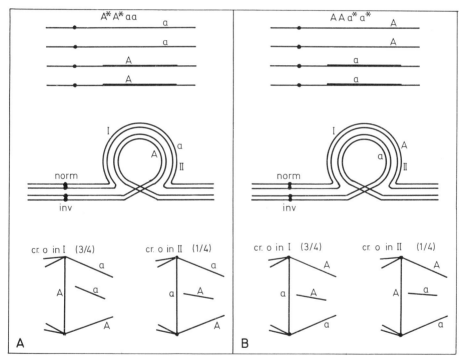

Fig. 3.15 A and B. The consequences of crossing-over in the inversion segment in a tetraploid paracentric inversion heterozygote, on the loss of a marker gene with the acentric fragment. All four chromosomes are shown at the top, but only two have been drawn for the demonstration of the course of events leading to marker loss. Since segment *I* is three times as long as segment *II*, the dominant allele will be lost in $\frac{1}{4}$ of the cases with a cross-over in the segment when it is in the inversion chromosome. It will be lost in $\frac{3}{4}$ of the cases when it is in the normal chromosome. In the tetraploid two pairing loops are present at a time, but only in the case of pairing between structurally different chromosomes. (Slightly altered after DOYLE, 1963)

quantitatively on the basis of segregations and bivalent frequencies alone unless information is available on the meiotic behavior of the second arm of the chromosome involved.

3.2.2.3 Autopolyploids with Translocations

The effect of translocations on pairing in autotetraploids is not analyzed as readily as that of inversions. No such simple phenomena as anaphase bridges are available. Even when all segments capable of pairing have a chiasma, the number of different configurations is large, and a detailed model is necessary for a quantitative comparison between random and preferential pairing. The first relatively complete model was constructed by LINNERT (1962), for estimating the expected segregation ratios in duplex heterozygotes in Oenothera. She neglected possible absence of chiasmata, which are of limited importance for segregations, but which may profoundly disturb an analysis of metaphase configuration frequencies. A model which takes chiasma failure into account was developed by SYBENGA (1973a) for a number of duplex translocation heterozygotes in autotetraploid rye. In Fig. 3.16 the eight chromosomes involved in the translocation are shown. In this model only the end segments are considered. Although interstitial chiasmata may to some extent complicate the situation, the simplification is considered acceptable for translocations in metacentric chromosomes with predominant distal chiasma formation.

There are two identical chromosomes of each type, but each segment is present four times. The four end segments have three possibilities of pairing. There are four different types of end segment and thus the possible number of combinations is $3^4 = 81$. These correspond to 81 configurations, several of which have the same appearance when the individual chromosomes cannot be distinguished at metaphase, as is usually the case. Six of these 81 have been drawn in Fig. 3.16, in alternate orientation. For each, the association of the specific arms has been indicated, and behind each the composition of the resulting gametes has been shown. A letter N indicates a normal complement, a letter T a translocation complement. Thus NT is a combination of a normal with a translocation complement in the resulting gamete, NN a combination of two normal complements and TT a combination of two translocation complements. The frequencies of being bound can be obtained from the diploid translocation heterozygote (2.2.4.2), but it should be noted that (a) the actual frequencies depend on the genotype, and (b) the conditions in the diploid may not completely reflect the conditions in the tetraploid, even when the genetic backgrounds are comparable. Only rough estimates are, therefore, available. Since only large deviations from 100% metaphase association have detectable effects, especially on the segregations, this need not be very disturbing. For this reason, in the example given, all non-translocation arms (c and d in Fig. 3.16) were assumed to be bound in all cases. The other arms have the possibility of homologous association ($a_1 - a_2$; $b_1 - b_2$; $a_3 - a_4$; $b_3 - b_4$) in which case they may also be assumed to associate without exception. Thus only the combinations $a_1 - a_3$;

a_1-a_4; a_2-a_3; a_2-a_4 and the same for b, associate with sufficiently lower frequencies to be worth considering.

Of the 81 configurations, six examples of which are given in Fig. 3.16, one consists of four homologously paired bivalents, eight consist of two bivalents and a quadrivalent, sixteen of two quadrivalents, sixteen of a bivalent and a hexavalent and forty of an octavalent. With pronounced preferential pairing between structurally homologous chromosomes, the category "four bivalents" will increase in frequency, as will bivalents in general. However, all configurations with a combination of segments of low chiasma frequency are liable to break down into smaller configurations, amongst which are bivalents. An increase in bivalent frequency as such is no proof of preferential pairing. Preferential pairing between single arms of metacentric chromosomes is not especially effective in causing an increase in ring-bivalent frequency, but it results in a decrease in the frequency of large configurations. When the total decrease exceeds that caused by multivalent break-down alone, preferential pairing may be concluded to have operated. In autotetraploids, bivalents and quadrivalents are formed in great numbers by the chromosomes not involved in the translocation, and therefore an increase in bivalent frequency is difficult to detect. The only way open is to estimate the decrease in configurations larger than quadrivalents. These are 65 out of 81 for the translocation complex, when all arms are bound and no preferential pairing operates. The octavalents are of two types: some have both translocated segments in heterologous association (i.e. for instance a_1-a_3, a_2-a_4 and b_1-b_3, b_2-b_4), others have only one in heterologous association (for instance, a_1-a_3, a_2-a_4 but b_1-b_2, b_3-b_4). The hexavalents always have either one or the other heterologously associated. A hexavalent of type 13 (Fig. 3.16) has segment a in heterologous association, but b homologous. For this type, at metaphase the hexavalent frequency is a^2 (rings)$+2a\cdot(1-a)$ (chains); the rest is quadrivalent or smaller. There are eight hexavalents of this constitution, and also eight in which b is paired heterologously. The octavalents are more complicated. Both translocated segments in heterologous pairing occur in 24 out of 40. The frequency of configurations larger than quadrivalent (from this group) then is a^2b^2 (octavalent rings)$+2a(1-a)b^2+2a^2b(1-b)$ (octavalent chains)$+(1-a)^2b^2+a^2(1-b)^2+2a(1-a)b(1-b)$ (hexavalents). The rest are quadrivalents or smaller configurations. Of the remaining 16 octavalents, eight have segment a heterologously paired, and eight, b. Here the frequencies of octavalents are a^2 (rings)$+2a(1-a)$ (chains) and b^2 (rings)$+2b(1-b)$ chains, respectively. Hexavalents are not formed, since the heterologous translocation segments are on opposite locations in the configuration: two quadrivalents may be formed (cf. Fig. 3.16), or an octavalent remains. The total number of configurations larger than quadrivalents can now be approximated as:

$$32a+32b+48ab+8a^2+8b^2-48ab^2-48a^2b+24a^2b^2. \qquad (3.8)$$

For $a=b=1$, this equals, as expected, 56.

For the example (SYBENGA, 1973a), the simplifications introduced (appr. 100% association of homologously paired arms, few interstitial chiasmata, diploid chiasma frequencies applicable for tetraploids) are acceptable. When they are not, the model must be extended. Even in this example the assumptions are not perfect: the finding of configurations consisting of five and seven chromosomes demonstrates that the homologously paired arms occasionally fail to have chiasmata. Some interstitial chiasmata also occur.

For three translocations in rye, the frequency of configurations larger than quadrivalents was scored (Table 3.14). In all cases this frequency was lower than expected, but this must be interpreted with caution. Especially in 273, one of the normal arms has a lower frequency of being bound than 1, especially in the tetraploid. This is true, to a lesser extent, also for 242. Taking this into account, it may be concluded that the frequency of large configurations is greatly reduced in 242, less so in 248, and little or not at all in 273. It is of interest to note that in the order 242, 248, 273 the location of one of the breakpoints is progressively further from the chromosome end. It is probable that the closer to the end the major segment of pairing initiation in situated, the more effective is the translocation in inducing preferential pairing.

In respect to segregation, the model contains an important simplification: the orientation is assumed to be alternate. When two or more independent fragments are formed, the different combinations of orientation are assumed to be formed at random (cf. 41 in Fig. 3.16). This has no effect on segregation when the fragments are symmetric (12 and 13 in Fig. 3.16). When however, asymmetric (derived) configurations orientate independently, a 50% unbalance may result, or in 50% of the meiocytes gametic types other than those expected with alternate orientation of the original configuration are formed (42 in Fig.

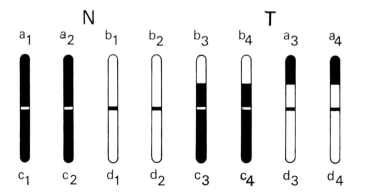

Fig. 3.16A and B. Autotetraploid with a reciprocal translocation (duplex heterozygote). (A) The eight chromosomes: two of each normal type and two of each translocation type. There are four different end segments (a, b, c, d) and each is represented four times (1, 2, 3, 4).
(B) Six of the 81 possible pairing combinations. Anaphase segregation in last column. N = normal complement; T = translocation complement. The first is the only type with exclusive bivalent formation. In all other combinations alternate orientation of the multivalent is assumed, but orientation is independent for separate multivalents. When two different examples for one combination are given, these are either different orientation types (41) or cases of break-up of larger configurations into smaller ones due to the absence of chiasmata. This may or may not result in different segregations (12, 13: no difference; 14, 42: different segregations). From SYBENGA (1973a)

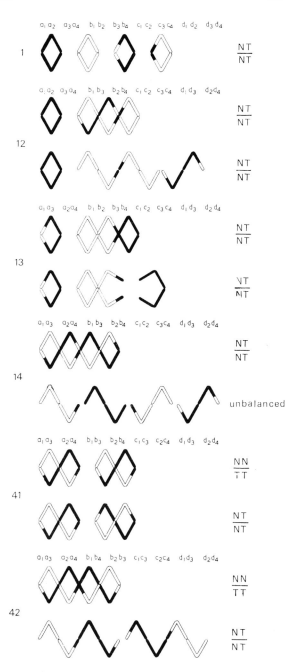

3.16). Thus, absence of chiasmata in critical segments affects the segregation ratios. Predominant alternate orientation is common in the example given, rye. The effect of deviations from alternate orientation can be minimized when the resulting unbalanced gametes can be filtered out. This is not normally possible in animals, but can be accomplished in plants when the segregating parent is the father: unbalanced pollen has considerably lower competitive capacities than balanced pollen. The correction thus obtained is not perfect but at least is an improvement. It was applied in the example given below.

Segregations were studied for translocations 240 and 248. The models required are complex and will not be considered in detail here. It appeared that within the species *Secale cereale* no deviation from random segregation could be demonstrated. This is at least in part due to the fact that in the large complex at most a single arm was affected, which has practically no effect on segregation. The second reason is that even with random pairing in tetraploid duplex translocation heterozygotes segregation is strongly "preferential", i.e. resembles that in an allotetraploid. This is partly a result of the fact that only a limited number of combinations of chromosomes gives viable progeny and, in plants, functional gametes. In addition, most of the gametes can be shown to combine a translocation set with a normal set of chromosomes (LINNERT, 1962; SYBENGA 1973a). In interspecific hybrids with the closely related genus *Secale montanum*, pronounced preferential pairing was observed, but it was uncertain to what extent this was due to the translocation, the species differentation, or their interaction. The use of translocations and other rearrangements in differentiating the originally identical genomes in autotetraploids, thus converting autotetraploids into functional allotetraploids, has been considered by BENDER and GAUL (1966) and, more extensively by SYBENGA (1969, 1973b).

In their analysis of meiosis in tetraploid $(4x=24)$ *Rhoeo discolor*, which has naturally occurring translocations involving all chromosomes, WALTERS and GERSTEL (1948) observed on the average 2.0 ring bivalents per cell (60 cells). In comparison with the diploid, the chiasma frequency was reduced by a factor 0.84. Interstitial chiasmata were scarce: 17 of 1,099 chiasmata were recorded as "multiple chiasmata". Because of the translocations, the chromosomes can best be represented by their ends, no two chromosomes being the same

Table 3.14. The frequencies of configurations larger than quadrivalents in three duplex translocation heterozygotes in autotetraploids of rye. The chromosomes have two translocated segments for each translocation with probabilities of having a chiasma a and b respectively. Assumed is: for 242: $a=0.1$ and $b=0.9$; for 248: $a=0.5$ and $b=0.75$; for 273: $a=0.8$ and $b=1$ (SYBENGA, 1973a)

Translocation	242	248	273
Configurations larger than quadriv.	14	36	13
Expected	23.9	56.0	17.1
Ratio obs./exp.	0.586	0.643	0.760
Expected when $a=b=1$	34.6	69.2	17.3
Number of cells	50	100	25

in the diploid: $a-b$; $b-c$; $c-d$; ...; $k-l$; $l-a$, 12 in all. In the tetraploid, all occur twice: a_1-b_1; a_2-b_2; b_3-c_1; b_4-c_2; ...; l_1-a_3; l_2-a_4, 24 together. A ring bivalent is formed when a_1-b_1 associates with a_2-b_2 etc. When a_1 pairs with a_2 in $\frac{1}{3}$ of the cases, and b_1 with b_2 in $\frac{1}{3}$, then the $a-b$ bivalent is formed with a frequency of $\frac{1}{9}$. Assuming that in the case of bivalent pairing the bivalent frequency of being bound (0.763) applies, a metaphase ring bivalent is formed in 0.763^2 of the bivalent pairing configurations. This makes an individual bivalent probability of $\frac{1}{9} \times 0.763^2 = 0.065$. For the twelve bivalents this is 0.78. The average of 2 observed is significantly higher, so that there is some preferential pairing, even though it is not strong.

3.2.3 Allopolyploids

In normal, stabilized allopolyploids exclusively bivalent pairing will occur between specific pairs of chromosomes. There is no essential difference with a diploid, and no analysis of variation in affinity between homeologous chromosomes is possible. There are two situations in which the differentiation is incomplete and where pairing between homeologues is (occasionally) possible: (a) new (natural or artifical) "amphidiploids" between species with an incomplete chromosomal differentiation and (b) special genotypes in established allopolyploids which (partially) remove mechanisms normally functioning in intensifying and completing the pairing differentiation between homeologues. The first situation is rather common, but the second is practically restricted to allopolyploid organisms which have been thoroughly analyzed cytogenetically, since it is requires an exceptional genotype.

3.2.3.1 New Allopolyploids (Amphidiploids)

There are numerous reports on pairing in new allopolyploids, usually studied in meiotic diakinesis-metaphase preparations. This often limits their value, since in several interspecific hybrids chiasma formation, even between completely homologous chromosomes as occur in the polyploid, may be incomplete. In others, however, chiasma formation may be complete, even after pairing between homeologues (RILEY, 1960). New amphidiploids may be expected to show a range of pairing differentiation between the chromosomes of the parental species from almost zero (as in an autopolyploid) to 1 in a perfect allopolyploid. SVED (1966) has formulated the expected quadrivalent frequencies for a model with two (terminal) pairing initiation points, when the differentiation between the chromosomes increases gradually (Fig. 3.17). At A, pairing may be A_1-A_1, A_2-A_2 (probability $1-a$) and A_1-A_2, A_1-A_2 (probability a). For B similar relations hold. If it is assumed that pairing is followed by chiasma formation, i.e. when no desynapsis occurs, and the average chiasma frequency is relatively high on both sides of any point of partner exchange, the relative frequencies of homologous bivalents, homeologous bivalents and quadrivalents at diakinesis-metaphase can be expressed in terms of a and b. Thus, homologous bivalents are formed with a frequency $(1-a)(1-b)$. With a gene constitution as in Fig.

3.17A, the gametes formed will *both* receive ͜   ͜ no segregation results. Homeologous bivalents will be formed with a   ͜ ͜ cy of $\frac{1}{2}ab$. The gametes resulting from such bivalents will show segreg͜   ͜ MM, $\frac{1}{2}Mm$ and $\frac{1}{4}mm$. The remaining

$$a+b-\tfrac{3}{2}ab \tag{3.9}$$

are quadrivalents. Here, with "random" chromos͜ ͜ ͜ ͜ junction at anaphase, $\frac{1}{6}mm$ gametes are expected. Thus the total frequenc͜ ͜ ͜ recessives is expected to be $\frac{1}{6}(a+b-\frac{3}{2}ab)+\frac{1}{4}(\frac{1}{2}ab)=$

$$\tfrac{1}{6}a+\tfrac{1}{6}b-\tfrac{1}{8}ab. \tag{3.10}$$

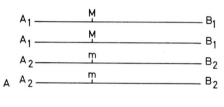

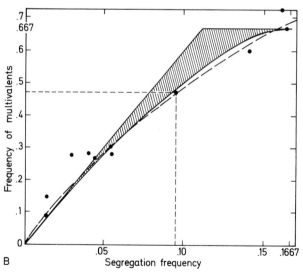

Fig. 3.17 A and B. The relation between multivalent frequency and gene segregation (gametes) in tetraploids with varying degrees of preferential pairing, when only the end segments initiate pairing and can do so independently, and when multivalent formation is not restricted by lack of chiasmata. (A) The chromosomes: two pairs of complete homologues (pair *1* and pair *2*, with arms *A* and *B*); varying degrees of differentiation between the pairs, one marker locus (*m*). (B) The relation. The right hand border of the shaded area gives the relation when both arms vary identically. The left hand border when one of the two arms consistently pairs preferentially. The points are examples of tetraploid Gossypium hybrids; the one marked is considered in the text. (After SVED, 1966)

This offers an opportunity to relate the frequency of quadrivalents (on the assumption of two independent pairing initiation sites) with genetic segregation ratios. Fig. 3.17B is from SVED (1966) and gives this relation, in addition to a set of observations on newly synthesized allopolyploids of Gossypium, from different authors and collected by PHILLIPS (1964b). The left border of the shaded area gives the line for $a=\frac{2}{3}$ with variable b (between 0 and $\frac{2}{3}$) or $b=\frac{2}{3}$ and variable a (0 to $\frac{2}{3}$). The right border gives the expected values for $a=b$, both between 0 and $\frac{2}{3}$. The observations (Table 3.15) fit this line well and thus permit an estimate of the average pairing differentiation between the chromosomes of each hybrid to be made. Taking, for instance, the hybrid $G.$ $hirsutum \times G.$

Table 3.15. Relationship between gene segregation and multivalent frequency in synthetic Gossypium allohexaploids $(2n=6x=78)$. When larger multivalents occurred (only in the first two hexaploids), those involving 7 or more chromosomes have been counted as 2 multivalents (PHILLIPS, 1964)

Hexaploid	Genomes	Segregation		Multivalents		
		No. loci	Av. segr.	Average per cell	per set	No. cells
G. hirsutum × arboreum	A$_2$ A$_h$ D$_h$	4	5.1:1	8.68	0.668	115
G. barbadense × arboreum	A$_2$ A$_b$ D$_b$	5	6.1:1	7.80	0.600	107
G. hirsutum × raimondii	A$_h$ D$_h$ D$_r$	10	9.5:1	6.16	0.474	108
G. hirsutum × harknessii	A$_h$ D$_h$ D$_{2-2}$	5	17.1:1	3.65	0.281	220
G. hirsutum × armourianum	A$_h$ D$_h$ D$_{2-1}$	5	17.4:1	3.96	0.305	258
G. hirsutum × aridum	A$_h$ D$_h$ D$_4$	4	21.3:1	3.48	0.268	279
G. hirsutum × lobatum	A$_h$ D$_h$ D$_7$	4	23.7:1	3.66	0.282	283
G. hirsutum × thurberi	A$_h$ D$_h$ D	3	32.9:1	3.61	0.278	44
G. barbadense × gossypiodes	A$_b$ D$_b$ D$_6$	4	71.6:1	1.13	0.087	109

raimondii, an average segregation ratio of 9.5:1 or 0.095 is combined with a multivalent frequency of 0.474. From Eqs. (3.9) and (3.10) a quadratic equation in a and b can be derived which, however, yields a negative discriminant. This is not unexpected, since the point (Fig. 3.17B) falls to the right of the shaded area. For the best estimates, a equals b. Then a separate value can be derived from the segregation ratio ($a=b=0.463$) and another from the multivalent frequency ($a=b=0.566$). Pairing apparently is in between that for a good autopolyploid and an allopolyploid. Since 10 loci are involved, most of the chromosomes are represented, and it is improbable that the difference between the two values is due to the unorthodox behavior of one or a few chromosomes. Since the model is only an approximation, there may be several reasons for the deviation. The method is in principle satisfactory for roughly estimating pairing differentiation in allopolyploids. The results for Gossypium indicate considerable variation in the affinities between different genomes. Since the points fit the curves reasonably well, there is little desynapsis.

3.2.3.2 Established Allopolyploids

Well balanced, established allopolyploids with exclusive and complete bivalent formation offer little opportunity for studying the pairing affinity relationships between their constituent genomes. Only when by some means the perfect differentiation can be broken down, is pairing between homeologous genomes to be studied. This has been done most extensively in hexaploid bread wheat (*Triticum aestivum*), but since the emphasis has generally been on the analysis of the factors affecting chromosomal differentiation rather than on the analysis of pairing itself, most reports are qualitative rather than quantitative in respect to pairing. From the numerous publications in the field, only a few can be considered, and the emphasis will be on those which present quantitative information on pairing.

A first indication that pairing differentiation between the genomes of wheat is enhanced by a controlling gene in chromosome 5B was the observation that in the absence of this chromosome pairing between homeologues was possible. SEARS and OKAMOTO (1958) found considerable trivalent and higher association in "diploid" wheat nullisomic for 5B, and RILEY and CHAPMAN (1958) reported bivalents and trivalents in haploid wheat from which this chromosome was lacking. It was soon found that this gene *Ph* (pairing homeologues) was located in the long arm and that it could be caused to mutate to less effective states artificially (RILEY et al., 1966). Some genotypes of *Aegilops speltoides* (Table 3.16) would repress the action of *Ph* (RILEY, 1960). Later, *Aegilops mutica* was found to have the same effect, but both *Ae. speltoides* and *Ae. mutica* had genotypes that failed to suppress the *Ph* gene of chromosome 5B (DOVER and RILEY, 1972a). Both species contain occasional supernumerary B-chromosomes which can take over the role of the *Ph* gene in the absence of 5B (VARDI and DOVER, 1972; DOVER and RILEY, 1972b). A similar effect of B-chromosomes enhancing the pairing differentiation between the chromosomes of related species was observed by EVANS and MACEFIELD (1973) in Lolium hybrids. Since no other source for the *Ph* gene in wheat has as yet been located it has been suggested that it might be derived from such a B-chromosome accidentally included in the original tetraploid amphidiploid. Detailed genetic analyses revealed that numerous genes in wheat, especially in the group 5 chromosomes but also in others, in some way or another affect homeologous pairing, and FELDMAN (1966) and FELDMAN et al. (1966) suggested that most of this action was through a regulation of somatic (premeiotic) pairing. In the absence of *Ph*, somatic pairing is relatively intense and brings both homologues and homeologues sufficiently near each other to enable them to pair effectively. The gene *Ph* suppresses somatic pairing to some extent, resulting in a less close proximity of homologous and homeologous chromosomes at the time of meiotic pairing. Only homologous chromosomes have sufficiently effective attraction systems to pair in the limited time available. An overdose of *Ph* restricts the premeiotic pairing even more, and results in such large distances between the chromosomes that even full homologues do not always manage to find each other. Then, when by accident homeologues are in the neighborhood, these may pair. The importance of premeiotic processes for pairing in wheat was also shown by

Table 3.16. Chromosome associations in different derivatives and hybrids of hexaploid wheat, *Triticum aestivum*. (From RILEY, 1960)

Type	No. cells	No. chrom.	Univ.	Open biv.	Ring biv.	III	Multivalents IV	V	VI
Normal haploids	750	$n=21$	18.59	1.15	0.01	0.03	—	—	—
Nulli V hapl.	218	$n-1=20$	7.55	2.63	1.19	1.50	0.06	0.01	0.01
Norm. dipl.	200	$2n=42$	0.14	20.93		—	—	—	—
Nulli V dipl.	60	$2n-2=40$	1.41	17.22		0.47	0.59	0.02	0.05
× *A. cylindrica*	40	$5x=35$	20.89	6.80		0.17	—	—	—
× *A. cyl.* (def. short arm V)	50	$5x=35$	19.62	7.18		0.34	—	—	—
× *A. cyl.* (def. long arm V)	50	$5x=35$	16.40	6.44		1.26	0.46	0.02	—
× *A. cyl.* (def. V)	100	$5x-1=34$	13.16	7.41		1.36	0.36	0.10	—
× *A. speltoides*	50	$4x=28$	6.04	6.64		1.38	0.76	0.04	—
× *A. speltoides*	50	$4x=28$	3.40	6.14		2.20	1.30	0.03	—
× *A. spelt.* def. V	30	$4x-1=27$	6.13	5.23		1.76	1.23	0.03	—
× *A. spelt.* def. V	30	$4x-1=27$	6.90	6.10		1.33	0.93	—	—
× *A. longissima*	100	$4x=28$	23.90	1.96		—	—	—	—
× *A. longissima* def. V	200	$4x-1=27$	8.65	7.02		0.79	0.49	—	—

the application of special treatments before meiosis and their effect on pairing (colchicine, DRISCOLL et al., 1967; high temperature, BAYLISS and RILEY, 1972b). The possibilities all this variation in the control of chromosome pairing offers for "chromosome engineering" in wheat are numerous and not merely theoretical. Several examples of practical application have been reported (SEARS, 1972a).

A quantitative analysis of the differential affinity between homeologues of wheat was performed by RILEY and CHAPMAN (1966), using a wheat-*Aegilops speltoides* hybrid (21 + 7 = 28 chromosomes) where the action of 5B on pairing restriction was suppressed. In the wheat variety used, there were 19 normal pairs of chromosomes, and two pairs of telocentrics. Of the marked chromosomes one arm was present in the normal dose, the other was absent. Several different combinations of marked chromosomes were combined with *Aegilops speltoides* and the double marked types studied. There were combinations of two marked homeologous chromosomes and of marked non-homeologous chromosomes (Fig. 3.18). The telocentrics participated in bivalents, trivalents and quadrivalents involving unmarked chromosomes or homeologous telocentrics, but never paired with a non-homeologous telocentric. For example, three homeologous combinations were constructed for the long arm of the group 5 chromosomes: 5A with 5B; 5A with 5D and 5B with 5D. Here the telocentrics paired with each other, most commonly in a rod-shaped bivalent, occasionally together with a normal chromosome in a triradial trivalent or with two normal chromosomes in a chain of four with the telocentrics at the ends (Fig. 3.18). When

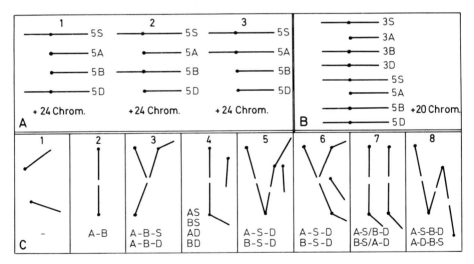

Fig. 3.18A–C. Estimating the relative pairing differentiation between the three genomes of allohexaploid wheat. (A) The chromosomes of the hybrid between *Triticum aestivum* (*n*=21), wheat and *Aegilops speltoides* (*n*=7). Two homeologous chromosomes 5 have been replaced by telocentrics. There are four homeologues, *A*, *B*, *D* and *S*; the remaining 24 chromosomes are irrelevant. In *1* the 5*A* and 5*B* chromosomes are marked; in *2*, 5*A* and 5*D*, and in *3*, 5*B* and 5*D*. (B) Two non-homeologous chromosomes replaced by telocentrics. (C) The different association types of the telos at MI (type of *A*₁). Which chromosomes have associated is shown; some configurations have more than one possible origin. (After RILEY and CHAPMAN, 1966)

the 5A long arm telocentric was combined with the 5B or 5D long arm telocentric, pairing between these telocentrics was relatively rare, but when 5B and 5D were combined, metaphase association between them was observed in more than half the cells scored. Apparently, 5B and 5D have a much closer affinity than either with 5A. Since quadrivalents involving both telocentrics could occur, pairing with the (unmarked) *Aegilops speltoides* homeologues must have been possible. Then there are six two-by-two combinations, and the frequency of each can be expressed as:

$$5A - 5B = a \qquad\qquad 5A - 5S = x$$
$$5A - 5D = b \qquad\qquad 5B - 5S = y$$
$$5B - 5D = c \qquad\qquad 5D - 5S = z$$

The few triradial trivalents were added to the telocentric bivalents to give the estimates of a, b and c as 0.08; 0.08 and 0.52 respectively (Table 3.17). The values for x, y and z can be derived indirectly as follows. Pairing between a telocentric and an unmarked chromosome can be of four types: when, for instance, a 5A and a 5B telocentric are present in addition to normal 5D and 5S chromosomes, there can be "marked" pairing between 5A and 5D, between 5A and 5S, between 5B and 5D or between 5B and 5S. The frequency T of these events can be expressed as $T_{(5A/5B)} = b + x + c + y$; $T_{(5A/5D)} = a + x + c + z$ and $T_{(5B/5D)} = a + y + b + z$. For each formula the frequency T equals the sum of the frequencies of marked and unmarked pairing (Table 3.17, columns (3), (8) and (11), the latter counting double since two telocentrics are involved). Thus $T_{(5A/5B)} = 0.01 + 0.54 + 2 \times 0.17 = 0.89$; $T_{(5A/5D)} = 0.81$ and $T_{(5B/5D)} = 0.46$. The values for a, b and c were obtained directly and can be substituted:

$$0.89 = 0.08 + x + 0.52 + y$$
$$0.81 = 0.08 + x + 0.52 + z$$
$$0.46 = 0.08 + y + 0.08 + z$$

Table 3.17. The frequencies of different types of association of telocentric chromosomes of homeologous group 5 at M I of meiosis in *T. aestivum* × *Ae. speltoides* hybrids (RILEY and CHAPMAN, 1966). Cf. Fig. 3.18

Combination of telocentrics	M I association type (Fig. 3.18)											Total cells
	Univ.	Telos together		sum	One telo/ One normal			sum	Two telos/ Two normal		sum	
Type in Fig. 3.18C	1	2	3	sum	4	5	6	sum	7	8	sum	
5A – 5B	21	7	1	8	48	4	2	54	17	—	17	100
5A – 5D	24	7	1	8	50	5	1	56	11	1	12	100
5B – 5D	14	44	8	52	30	—	—	30	3	1	4	100
Column	(1)	(2)	(3)	(4)	(5)	(6)	(7)	(8)	(9)	(10)	(11)	(12)

which leads to $x=0.10$; $y=0.19$ and $z=0.11$. The relative affinities between the four chromosomes for homeologous group 5, long arm, can now be expressed as:

5A−5B	0.08	5A−5S	0.10
5A−5D	0.08	5B−5S	0.19
5B−5D	0.52	5D−5S	0.11

These indicate considerable variation in affinity between the genomes, and it is somewhat surprising that under the conditions of relaxed pairing barrier, the supposedly homologous B- and S genomes pair in fact less effectively than the B and D genomes. Also noteworthy is this close affinity in view of the fact that nulli-tetra compensation is hardly better between the B and D genomes than between the A and B or D genomes. Apparently, species differentiation is expressed differently for different parameters.

The above mentioned affinities are not to be compared directly with homologous pairing. In a similar wheat—*Ae. speltoides* hybrid with normal chromosomes 5 but 6B replaced by two fully homologous telocentrics of one of the arms, RILEY and CHAPMAN (1966) observed metaphase association between these telocentrics with a frequency of 0.86. If this is a representative value, the 0.52 frequency of 5B−5D association in the hybrid would represent a 60% of normal (homologous) affinity.

The affinity between homeologous chromosomes of different genomes can be analyzed less specifically, not using marked chromosomes, by a method developed by GAUL (1959). The purpose, then, is no more than determining the number of pairs of chromosomes that are in principle capable of (pairing and) forming chiasmata. This may be of interest, for instance, in hybrids between two allopolyploids, combining several more or less related genomes, and where partial desynapsis may inhibit the complete expression of homologies. In such a situation, the number p of chromosomes capable of association is given as

$$p = \frac{X^2+X-B}{(2X-B)C}$$

in which the total number of chiasmata is X, the total number of paired chromosomes is B and the number of cells analyzed is C. A quadrivalent is equivalent to two bivalents, etc. Applying this formula to his observations on hybrids between wheat (*Triticum aestivum*, $2n=42$) and *Agropyron intermedium* ($2n=42$), in spite of a relatively low frequency of bivalents, GAUL (1959) could conclude that in principle each of the three genomes of wheat has enough homology with one of the Agropyron genomes to make (occasional) association possible. The analysis could have been carried further if the distribution of bivalents over cells had also been recorded.

3.2.3.3 Secondary Association

Remnants of homeologous association not expressed in bivalents and multivalents where the opportunities for effective pairing resulting in chiasma formation were too restricted, can be found in *secondary association* of homeologous bivalents. These are then closer together than non-related bivalents. With some exceptions, the descriptions of such associations tend to be qualitative and vague and sometimes are mere suggestions, especially when the different bivalents in the cell are too similar to be recognized. There are a number of examples of rather exact analysis of secondary association of recognizable bivalents, differing from other bivalents either by natural characteristics or by being marked. KEMPANNA and RILEY (1964) recorded the relative location in the metaphase plate of wheat, of bivalents marked by replacing one member of each of two chromosome pairs by a telocentric. This resulted in readily recognizable heteromorphic rod bivalents (Fig. 3.19A). Different combinations of marked homeologous and of marked unrelated bivalents were used. The relative positions of the two marked bivalents on squashed, and consequently linear first metaphase plates were expressed in terms of the numbers of (unmarked) bivalents by which they were separated. First, the position of single marked bivalents in the plate was tested. The 21 possible positions of the marked bivalent were divided into two parts each of ten positions, numbered 1 to 10 counting inwards from each extremity of the cell, and the central position was 11. Provided that the distribution along the plate was random, the probabilities of the marked bivalent occurring in positions 1 to 10 must be equal and twice those of its occurring in position 11. In 50 cells from each of 3 plants the marked bivalent appeared to be located entirely at random. Next, the number of bivalents separating two marked bivalents was compared with the number expected when both were independently and at random located on the equator. The expectation was calculated from the equation

$$\frac{2(n-1-r)}{n(n-1)} \tag{3.11}$$

where n is the total number of bivalents and r is the number of intervening bivalents. The random distribution is a straight line which in the example of wheat, where $n=21$ and $r=0-19$, runs from a frequency of 0.0952 (no intervening bivalents) to 0.0048 when there are 19. When comparing the observed frequencies for homeologous bivalents (five different combinations pooled) and non-homeologous bivalents (nine different combinations pooled) with the random expectation, a very significant deviation was found (Fig. 3.19B). Between them, however, homeologous and non-homeologous marked bivalents also deviated significantly in respect to the distribution of the number of intervening bivalents. Both had an excess of high numbers but the homeologous marked bivalents had a great excess of the 0 and 1 class of numbers of interveners, combined with a deficit in the intermediate classes. The excess of high numbers cannot be explained, but the excess of the 0 class for homeologous bivalents clearly points to secondary association.

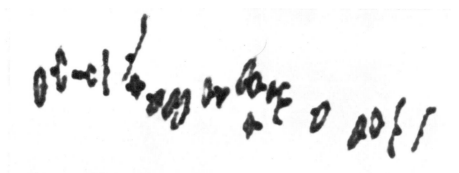

A

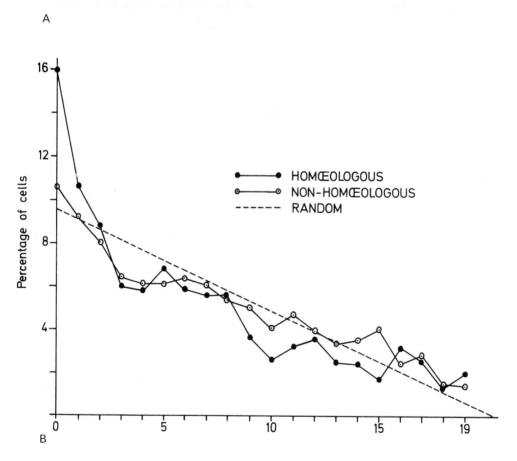

B

Fig. 3.19A and B. The analysis of secondary association on the basis of the location of two marked bivalents in a linear M I plate. (A) *Triticum aestivum* ($n = 21$); two homeologous bivalents are marked, each consisting of one normal and one telocentric chromosome: there are 15 intervening bivalents (Photograph courtesy R. RILEY). (B) The distribution of the number of intervening bivalents between two marked homeologous, and between two non-homeologous bivalents. The dashed line is the expected (random) distribution. There is a significant excess of secondarily associated homeologous bivalents. (After KEM-PANNA and RILEY, 1964)

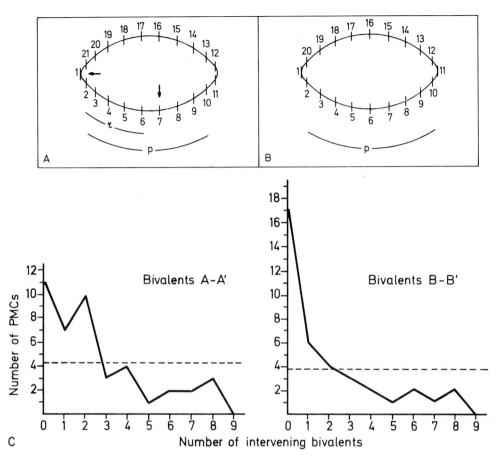

Fig. 3.20 A–C. The analysis of secondary association between recognizable bivalents in a circular M I plate. (A) Uneven number of bivalents (here $n=21$). The (lowest) number of intervening bivalents equals r (see example indicated by arrows); $\dfrac{n-1}{2} = p$. (B) Even number of bivalents (here $n=20$); $\dfrac{n}{2} = p$. (C) Two examples of recognizable bivalent pairs in *Aegilops triaristata*. The expected number of cells with any number of intervening bivalents is shown by the dashed line: it is equal for each number and depends on the number of cells studied for the particular pair of bivalents. (After LACADENA and PUERTAS, 1969)

The detection of secondary association with this method suffers from a few inevitable defects: the linear metaphase plate has been derived from a two dimensional one by squashing, which may have altered the relative positions of the bivalents. In addition, by producing a one-dimensional plate, part of the information contained in the original, two-dimensional plate is lost. Secondly since telocentric (half-)chromosomes are used to mark the bivalents, less chromosomal material is available for secondary attraction than with complete bivalents.

These two disadvantages may at least partially be avoided when the metaphase plate can be studied in a less disturbed state and when normal bivalents can be recognized. This was realized to a considerable degree in the *Aegilops triaristata* material analyzed by LACADENA and PUERTAS (1969). It is an allohexaploid $(2n=6x=42)$ related, although not very closely, to wheat. Allotetraploid forms occur also in this species, and have the genome constitution $C^u C^u M^t M^t$. In the hexaploid $(C^u C^u M^t M^t M^{t_2} M^{t_2})$, the M^t and M^{t_2} genomes are chromosome-morphologically similar, with distinct differences between several of the individual chromosomes, even at metaphase of meiosis. This makes it possible to recognize homoeologous bivalents without artificial markers. There is perfect bivalent pairing, presumably exclusively between complete homologues, and the authors were interested in estimating the degree of secondary association. It appeared that the metaphase arrangement was approximately circular with the bivalents at the perifery, and only those cells were analyzed in which the circle could still be recognized after squashing. Although not all disturbing factors will have been avoided, a better analysis is possible than in a linear plate. In a circle (Fig. 3.20A, B) the number of intervening bivalents, of course, can be expressed in two ways around the circle, but only the smaller distance was considered. Now, the expected frequency is the same for each number of intervening bivalents.

In the case of $n=21$, there are $\dfrac{n-1}{2}=10$ classes with numbers ranging from

0 to 9. LACADENA and PUERTAS (1969) recorded the numbers of intervening bi-valents for the different pairs of homoeologous bivalents in 30 to 54 circular metaphase plates and compared these with the expected frequencies (Fig. 3.20C, D). There was some difference between the different chromosomes, but a large and very significant excess of the low frequencies of intervening bivalents was universally observed, but (again) a slight increase in the higher numbers. Such an increase can be explained when one considers that opposite positions in a ring may well be occupied by bivalents that before prometaphase movement were very close together. A derivation of the average number of intervening bivalents was not given by LACADENA and PUERTAS (1969) and is somewhat different for even and odd numbers of bivalents. When n is odd (Fig. 3.20A), for any position of one of two bivalents, the second one has $(n-1)$ possibilities.

The maximum number of intervening bivalents (r_{max}) is $\dfrac{n-1}{2}-1$. There are

two positions with this number. There are also two positions with the next

largest number of intervening bivalents: $r=\dfrac{n-1}{2}-2$, etc. For the nearest positions

$r=0=\dfrac{n-1}{2}-\dfrac{n-1}{2}$. The sum is

$$\sum_{\left(x=\frac{n-1}{2}\right)}^{1} 2\left\{\frac{n-1}{2}-x\right\}=2\times\left(\frac{n-1}{2}\right)^2-2\times\left(1+2+...\frac{n-1}{2}\right).$$

For $n-1$ positions, the average then is

$$\bar{r}=\frac{n-3}{4}. \qquad\qquad\qquad\qquad\qquad\qquad\qquad\qquad (3.12)$$

For $n=21$ as in *Aegilops triaristata* the average number of intervening bivalents is $\frac{18}{4}=4.5$. The observed average for the A and A′ bivalents was 2.49 (43 cells) and for the B and B′ 1.87 (cf. Fig. 3.20C, D). This can also be derived more directly: There are 10 classes of numbers of intervening bivalents (0–9), all with equal probability. The sum is $1+2+3\ldots+9=45$. The average number of intervening bivalents then equals $\frac{45}{10}=4.5$. The observed average was 2.49 (in 43 cells) for the A and A′ bivalents. It was 1.87 for the B and B′ bivalents (compare Fig. 3.20C, D). The difference is large, and significant in a χ^2 test, for both cases. Although LACADENA and PUERTAS (1969) do not consider the variances, they are worth analyzing. The data presented permit an estimate: for the A−A′ bivalents the variance is 6.40; for the B−B′ bivalents it is 6.01. For the expected variance the following formula can be derived (STAM, 1974):

$$\sigma^2 = \frac{n(n^3-2n^2+2n-1)}{3(2n-1)^2}. \tag{3.13}$$

For both bivalent sets this leads to a value of 4.0. A comparison of expected and observed variances is of interest when the mean does not deviate from expected in cases where the peaks at high and low numbers of intervening bivalents compensate each other.

For even values of n, there are again $n-1$ positions for the second bivalent, but now there is only one for the maximum number of intervening bivalents, $r_{max}=\frac{n-2}{2}$. For the others there are two positions with $r=\frac{n-2}{2}-1, r=\frac{n-2}{2}-2,$ $\ldots, r=0=\frac{n-2}{2}-\frac{n-2}{2}$. Now, the sum is $2\times\left(\frac{n-2}{2}\right)^2-2\times\left(1+2+\ldots\frac{n-2}{2}\right)+\frac{n-2}{2}$ and the mean over $n-1$ positions:

$$\bar{r} = \frac{(n-2)^2}{4(n-1)}. \tag{3.14}$$

When $n=20$, as, for instance, for a nullisomic of a normally $n=21$ plant, the average would be $\dfrac{18^2}{4\times 19}=4.27$.

A third type of analysis, for the case the bivalents are randomly distributed in a two-dimensional circular area, secondary association can be analyzed along lines similar to those followed by FELDMAN et al. (1966) for somatic pairing. The equation used is a standard probability density function for a circle with radius 0.50:

$$f X^{(\chi)} = \frac{16\chi}{\pi}\left[\cos^{-1}\chi - \chi(1-\chi^2)^{1/2}\right] \tag{3.15}$$

where X is the distance between two randomly distributed points in a circular area.

Chapter 4. The Analysis of Distribution: Centromere Coorientation

Although distribution is the only one of the three central elements of meiosis generally accessible to direct analysis, it has not received nearly as much attention in the literature as chromosome pairing and crossing-over. A major reason is undoubtedly the fact that relatively little variation from randomness is observed in chromosomes not mechanically connected, especially in diploids. A few exceptions have been noted, with indications of "affinity" and "distributive pairing". Lack of satisfactory markers is a second reason. Univalent segregation in haploids and species hybrids, and trivalent segregation in triploids has been studied somewhat more frequently, and sometimes deviations from random distribution systems can be quantitatively described, if not explained. More details are available on the segregation of chromosomes associated in multivalents, but even here the complexity of the phenomena makes it difficult to design satisfactory models on the basis of which quantitative interpretations of the observations can be given.

4.1 Orientation and Reorientation in Bivalents; Independence and Interdependence

The stage primarily responsible for the distribution of the chromosomes at meiosis (segregation) is first anaphase (A I). The second anaphase (A II) is a separation of the two units of the double centromeres and closely resembles mitosis. Unlike in mitosis, the chromatids attached to the same centromeres at A II need not be genetically identical, but their distribution to the two poles is essentially a simple process. As far as it has been studied, it appears to be random, with one chromatid passing to each pole, and independent for the different chromosomes, except in exceptional cases of meiotic drive and in some structurally abnormal material. First anaphase distribution, on the other hand, is subject to several variables, even when meiotic drive and chromosome structure are not the immediate causes of irregularities. It depends on the previous coorientation of the unsplit centromeres of the associated chromosomes in the metaphase configurations, which may have specific characteristics affecting orientation. The simultaneous orientation of two centromere halves of a chromosome to the same poles is called syntelic and is normally restricted to A I. The orientation of the two centromere halves to different poles is amphitelic and occurs at A II. Occasionally an A I syntelic orientation may be converted into amphitelic orientation (BAUER et al., 1961).

The observation of orientation and subsequent segregation is usually more direct than that of pairing and crossing-over, although there may be special complications. For instance, in many mammalian preparations of meiosis, undisturbed metaphases are scarce. Or the final (critical) orientation may not be attained before the very end of metaphase, and observations made earlier during metaphase cannot be relied upon as a source of information on segregation. In such cases occasionally the segregation of marked chromosomes as observed at the second meiotic division or even later may be used, but a sufficient number of recognizable chromosomes is only infrequently available for a detailed analysis of segregation.

In most higher organisms the chromosomes normally have one centromere. In some animals, notably insects of the order of the Hemiptera and a few, mostly monocotyledonous plants (Juncales, Cyperales, some dicotyledonous Ranales) the chromosomes are *holokinetic*, i.e. have centromere activity distributed all over the chromosome. Others (some nematodes and scorpions) have a similar condition but probably with more distinct centromeres located at numerous places along the chromosomes (*polycentry*). The two conditions cannot readily be distinguished. Both require special mechanisms for segregation, but their complex behavior will not be considered here.

In general, orientation of a centromere towards a pole remains stable only when a counter-force is exerted (NICKLAS, 1968, 1971; SYBENGA, 1972a). This pull is normally produced by the centromere of another (homologous) chromosome attached to the first one by means of a chiasma. When for the some reason or another the pull from one centromere lapses, the other is able to reorientate and for a new orientation may choose the other pole. The second centromere can then move towards the opposite pole. As long as both centromeres continue to pull, the new orientation remains stable. By micromanipulating two rod-shaped bivalents of *Melanoplus differentialis*, HENDERSON and KOCH (1970) induced them to interlock in such a way that the two centromeres of one bivalent orientated on one pole, the two centromeres of the second bivalent orientating on the other. The condition remained stable for a considerable period of time.

In general it is assumed (on sound grounds) that different bivalents are independent in respect to their orientation as long as no physical connection between them is apparent. If there is, they may be considered multivalents, which will be considered in Section 4.2. There are some notable exceptions. Already in 1916, PAYNE concluded that in mole cricket males (*Gryllotalpa* spp.) in the absence of any physical connection the X chromosome would always orient to the same pole as the large chromosome of a heteromorphic bivalent in the same cell (cf. WHITE, 1954, 1973). This behavior was confirmed by CAMENZIND and NICKLAS (1968) who changed the orientation of the X and the heteromorphic bivalent respectively by micromanipulation. Artificial reorientation of the heteromorphic bivalent made the X-chromosome follow. Reorientation of the X, however, did not affect the heteromorphic bivalent, but the X returned to its original orientation. Removal of the heteromorphic bivalent caused the X-chromosome to continue moving from one pole to the other. Comparable micromanipulation on sex chromosomes and autosomes in Crane fly spermatocytes (FORER and KOCH, 1973) revealed interdependence in respect to the orientation of these chromosomes. Interdependent behavior in respect to segregation must lead to genetic linkage between the chromosomes that move to the same pole. Such linkage between different autosomes has been described by WALLACE (1961) for special strains of mice, but without cytological confirmation. It was termed *affinity*, and apparently non-homologous chromosomes from one parent tended to go together to the same pole, those of the other parent moving to the other pole.

A *quantitative meiotic analysis* of interdependent orientation can be made only when the chromosomes of one parent can be distinguished from those of the other. This is possible in hybrids between species with chromosome size differences which are maintained in meiosis of the hybrid, and were at the same time extensive pairing and chiasma formation remains possible. Such hybrids have been described for instance in the genera Gossypium (cotton) and Lolium (ryegrass). In some cotton hybrids (PHILLIPS, 1964) the parental types appeared preferentially in F_2, suggesting affinity, but this resulted from gametic and zygotic selection, rather than from preferential segregation. In others, segregation was random. In two hybrids with conspicuously different chromosomes a possible correlation in orientation between the heteromorphic bivalents was studied. Only the bivalents were recorded, neglecting the behavior of univalents that moved to the poles. The hybrid $(2n=26)$ of *Gossypium sturtii* (C-genome) with *G. lobatum* (D-genome) had on an average only 5.33 bivalents, ranging from 2 to 11. The distribution of the numbers of bivalents per cell is given in Table 4.1. All theoretically possible combinations of orientation are shown for each number of bivalents separately, together with the number of cells found for each of these combinations. Comparison with the corresponding (folded) binominal distributions failed to show deviation from randomness. In another hybrid (*G. sturtii* × *G. aridum*) the average number of bivalents was 8.98 with a range of 2–13 (the maximum). Here again the orientation was random. A very similar analysis of heteromorphic bivalents (Fig. 4.1) in Lolium hybrids (REES and JONES, 1967) also failed to show correlated coorientation, and the same was true for a *Nicotiana glutinosa* × *N. otophora* hybrid (SFICAS, 1963).

Correlated orientation between trivalents in autotriploids, even when not heteromorphic, can be studied either at metaphase by recording the pole towards which the single chromosome moves, or by analyzing anaphase segregations. SATINA and BLAKESLEE (1937) compared the anaphase chromosome distribution in PMCs of triploid Jimson weed (Datura) with a binomial distribution (Table 4.2) and noted an excess of the extremes. This can be interpreted to mean either that the trivalents tended to assume a corresponding orientation or that the univalents accompanying bivalent association tended to move to the same pole. Without metaphase studies no distinction between these two possibilities can be made.

There is an enormous amount of *qualitative* descriptive information on orientation of pairs of chromosomes, especially of the sex chromosomes. Whereas in general some form of attachment (by chiasmata or otherwise) is required, there are numerous instances where coorientation without attachment or even without prior pairing, at most after a short meeting at metaphase (touch-and-go

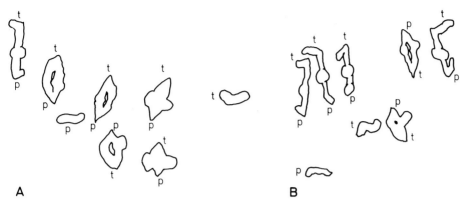

Fig. 4.1 A and B. Examples of orientation of the chromosomes of *Lolium perenne* (*p*) and *L. temulentum* (*t*) in the heteromorphic bivalents of the F$_1$ at M I. The difference in size is maintained in the hybrid, and sufficient for recognition. In (A), five of the six orientate to the same pole. In (B), four orientate to one pole and two to the other. There are two univalents in both examples. Drawn in outline after photographs by REES and JONES (1967)

Table 4.1. The distribution of *Gossypium sturtii* (C genome) and *G. lobatum* (D genome) chromosomes to the poles at the first meiotic division in the hybrid. Average number of bivalents 5.33. For each number of bivalents the observed numbers of cells with all possible distributions are recorded. The χ^2 values refer to a comparison with a binomial distribution (PHILLIPS, 1964a)

Bivalents per cell	No. of cells	Distributions (folded)							χ^2	*d.f.*	*P*
2	12	1-1 8	2-0 4						1.33	1	0.25
3	22	2-1 18	3-0 4						0.55	1	0.47
4	36	2-2 13	3-1 22	4-0 1					0.30	1	0.61
5	39	3-2 25	4-1 11	5-0 3					0.42	1	0.51
6	39	3-3 14	4-2 19	5-1 5	6-0 1				1.30	2	0.55
7	23	4-3 16	5-2 5	6-1 2	7-0				2.04	1	0.17
8	25	4-4 5	5-3 12	6-2 6	7-1 2	8-0			0.67	2	0.73
9	3	5-4 2	6-3 1	7-2	8-1	9-0					
10	3	5-5 2	6-4 1	7-3	8-2	9-1	10-0				
11	1	6-5 1	7-4	8-3	9-2	10-1	11-0				
	203										

Table 4.2. Anaphase I chromosome distribution in 1,000 PMCs of triploid Datura ($3n=36$) compared with a binomial distribution (SATINA and BLAKESLEE, 1937)

Distribution:	12-24	13-23	14-22	15-21	16-20	17-19	18-18
% observed	0.8	4.5	8.5	14.5	22.9	30.8	18.0
% expected	0.05	0.6	3.2	10.7	24.2	38.7	22.6

mechanism) appears to be possible (SCHRADER, 1953; WHITE, 1954, 1973; DARLINGTON, 1937, 1965; JOHN and LEWIS, 1965). The phenomenon is scarce in plants but has a parallel in the behavior of the two (different) telocentrics of the same metacentric chromosome which may coorient without apparent connection. Only the centromeric regions are homologous. STEINITZ-SEARS (1963) found coorientation between two such telocentrics in wheat in about half the number of cells analyzed, when the centromeric region was large enough to behave normally in respect to movement. If less stable centromeres were involved, the degree of coorientation was lower. It should be noted that such centromeres preceed the rest of the chromosome in anaphase movement, and in this case, therefore, coorientation must have taken place *before* movement. In such cases, the chromosomes are in fact univalent but at least have some clearly homologous segments. Univalents without any homology (although the presence of incidental duplications cannot be excluded) can also coorientate in *distributive pairing* (GRELL, 1964), which apparently takes place long after homologous pairing and probably during metaphase. It is primarily a function of size: only chromosomes of similar size associate sufficiently effectively to affect each other's orientation mechanism. Distributive pairing has been studied most extensively in Drosophila and almost exclusively by genetic methods. GRELL and DAY (1970) reported associations of non-homologues, of similar size in oogonial cells of *Drosophila melanogaster*.

Cytological analysis of univalent segregation has further concentrated on X0 systems of sex determination in animals and on trisomics, haploids, interspecific hybrids and asynaptics. The single X-chromosome in X0 males can be seen to move first from one pole to the other until at the stage when the autosomes have collected properly at the metaphase plate, it selects a definite pole to move to. The centromere is then pointed toward this selected pole, but the chromosome body moves towards this pole only slightly ahead of the separating autosomes at anaphase. (NICKLAS, 1961; WHITE, 1973).

In haploids etc. large numbers of univalents can be seen either to lag at the equator or to move to the poles in a more or less organized manner. The most simple *quantitative* test on univalent segregation is a comparison of the distribution of the numbers of univalents at either pole with a binomial distribution. BELLING and BLAKESLEE'S (1927) data on haploid *Datura stramonium* (Jimson weed) are a good example (Table 4.3) and show an excellent fit. Although for some reason or another all univalents managed to move to one of the poles (in the cells analyzed), this does not seem to involve any coorientation. In that case more equal numbers of chromosomes would be expected in the two poles, resulting in a narrower distribution about the mean. With univalents

Table 4.3. Anaphase I chromosome distribution in 100 PMCs of haploid Datura ($n=12$) compared with a binomial distribution (BELLING and BLAKESLEE, 1927)

Distribution:	1-11	2-10	3-9	4-8	5-7	6-6
% observed	1	3	12	19	38	27
% expected	1	3	11	24	39	23

lagging at the equator, the analysis is more complicated, and either cells without laggards must be analyzed, or the cells with laggards must be separated into classes with specific numbers of laggards, and the classes analyzed separately. Whenever bivalents are formed, their number must be subtracted from the chromosome number recorded at anaphase, as they would affect the segregation in a directed way and thus disturb the analysis (RILEY and CHAPMAN, 1957). This requires separate metaphase and anaphase observations of the same material, which have usually not been carried out. This makes many reports on anaphase segregation inaccessible to quantitative analysis.

A more sophisticated quantitative analysis of anaphase segregation was designed by SFICAS (1963) and applied to interspecific Nicotiana hybrids with variable chromosome pairing, and to a Nicotiana haploid. The segregation was recorded at the time the bivalents were still intact, and only those cells were used in which the univalent distribution to the poles could be clearly identified. With a total of N chromosomes and i bivalents, the total univalent frequency is $u=N-2i$. When x univalents go to one pole and $u-x$ to the other, the distribution of chromosomes to the poles is given as

$$D_{i+x}= 2\binom{u}{x}(1/2)^u \text{ for } x<\frac{u}{2}$$
$$\text{and } = \binom{u}{x}(1/2)^u \text{ for } x=\frac{u}{2}. \tag{4.1}$$

This equation applies when the number of bivalents is constant for all cells considered. As an approximation the average number of bivalents can be taken, but it is better to use more exact formulae. Let the probability that i bivalents are formed be B_i ($i=0, 1, 2 \ldots N/2$, when N is an even number, and $i=0, 1, 2,\ldots, (N-1)/2$ when N is an odd number). Then the probability that all chromosomes go to one pole and none to the other is only realized when there are no bivalents (B_0) or $D(0,N)=2(1/2)^N B_0$. One chromosome to one pole and all but one to the other can be realized when there is a bivalent (but then all univalents go to the same pole) and also when there is no bivalent, but the univalent distribution is 1 vs. $N-1$. Therefore,

$$D(1,N-1)= 2\binom{N}{1}(1/2)^N B_0+ 2\binom{N-2}{0}(1/2)^{N-2} B_1.$$

This can be generalized for the distribution y vs. $N-y$ when $y<N/2$:

$$D(y,N-y)= 2\binom{N}{y}(1/2)^N B_0+ 2\binom{N-2}{y-2}(1/2)^{N-2} B_1+\ldots$$
$$+ 2\binom{N-2y}{0}(1/2)^{N-2y} B_y. \tag{4.2}$$

When N is an even number, and $y = N/2$ there is also

$$D\left(\frac{N}{2}, \frac{N}{2}\right) = \binom{N}{y}(1/2)^N B_0 + \binom{N-2}{y-2}(1/2)^{N-2} B_1 + \dots$$
$$+ \binom{N-2y}{0}(1/2)^{N-2y} B_y. \tag{4.3}$$

The application of these equations is greatly simplified by using the tables supplied by SFICAS (1963) for the coefficients of B (Table 4.4). For example, for N being an even number, when $N=2$ (lower right end of Table 4.4A)

$D(0,2) = 0.5000 B_0$
$D(1,1) = 0.5000 B_0 + 1.0000 B_1.$

When $N=4$:

$D(0,4) = 0.1250 B_0$
$D(1,3) = 0.5000 B_0 + 0.5000 B_1$
$D(2,2) = 0.3750 B_0 + 0.5000 B_1 + 1.0000 B_2.$

It may be noted that necessarily $B_0 + B_1 + B_2 = 1$, for $N=4$, and when $B_0 = B_1 = 0$, $B_2 = 1$, i.e. there are two bivalents and the distribution is (2,2) without exception. For odd numbers of N, Table 4.4B is used, and when $N=3$:

$D(0,3) = 0.2500 B_0$
$D(1,2) = 0.7500 B_0 + 1.0000 B_1.$

For $N=5$:

$D(0,5) = 0.0624 B_0$
$D(1,4) = 0.3124 B_0 + 0.2500 B_1$
$D(2,3) = 0.6250 B_0 + 0.7500 B_1 + 1.0000 B_2.$

These distributions are based on random segregation of univalents, and can be compared with the observed distributions using a χ^2 test. With large numbers of univalents, another statistical approach is possible, using the second moment of the distribution curves, which gives an estimate of kurtosis (MATHER, 1965). A peaked distribution will have a smaller second moment, and a flattened curve a higher second moment.

Along similar lines as followed for taking account of the bivalents, lagging univalents can be taken care of, but this makes the equations somewhat more complicated, and larger numbers of observations are required for the same resolution of the analysis. It may then be simpler to group the observations into classes with the same number of laggards.

The same analysis is possible in triploids, when a trivalent is considered as a bivalent with a univalent, since the third chromosome may move at random to either pole. In tetraploids, univalents and trivalents can be treated in the same way, while quadrivalents are counted as two bivalents.

SFICAS (1963) applied his analysis to a number of interspecific Nicotiana hybrids with $N=24$ and $N=36$ and a polyhaploid of *N. tabacum* with $N=24$, with one chromosome substituted by a *N. plumbaginifolia* chromosome. The segregations had earlier been reported by SFICAS and GERSTEL (1962). A selection of these data is given in Table 4.5. Most interspecific hybrids had peaked distributions, i.e. more equal than expected. This suggests some coorientation in spite

Table 4.4A and B. Coefficients for calculating the distribution of univalent chromosomes to the poles. (A) for N being an even number ≤24; (B) for N being an odd number ≤19 (SFICAS, 1963)

(A)

0.0000	0.0000	0.0002	0.0013	0.0051	0.0160	0.0413	0.0877	0.1559	0.2338	0.2976	0.1612
0.0000	0.0000	0.0001	0.0007	0.0035	0.0126	0.0356	0.0813	0.1525	0.2372	0.3083	0.1682
	0.0000	0.0000	0.0004	0.0022	0.0092	0.0296	0.0739	0.1479	0.2403	0.3204	0.1762
		0.0000	0.0001	0.0012	0.0062	0.0233	0.0654	0.1416	0.2428	0.3338	0.1855
			0.0000	0.0005	0.0037	0.0171	0.0555	0.1333	0.2444	0.3491	0.1964
				0.0001	0.0017	0.0111	0.0444	0.1222	0.2444	0.3666	0.2095
					0.0005	0.0059	0.0322	0.1074	0.2417	0.3867	0.2256
						0.0020	0.0195	0.0879	0.2344	0.4102	0.2461
							0.0078	0.0625	0.2187	0.4375	0.2734
								0.0312	0.1875	0.4687	0.3125
									0.1250	0.5000	0.3750
										0.5000	0.5000
											1.0000

(B)

0.0000	0.0000	0.0006	0.0036	0.0148	0.0444	0.1036	0.1922	0.2884	0.3524
	0.0000	0.0002	0.0020	0.0104	0.0364	0.0944	0.1888	0.2968	0.3710
		0.0000	0.0010	0.0064	0.0278	0.0834	0.1832	0.3054	0.3928
			0.0002	0.0032	0.0190	0.0698	0.1746	0.3142	0.4190
				0.0010	0.0108	0.0538	0.1612	0.3222	0.4512
					0.0040	0.0352	0.1406	0.3282	0.4922
						0.0156	0.1094	0.3282	0.5468
							0.0624	0.3124	0.6250
								0.2500	0.7500
									1.0000

Table 4.5. The distribution of the chromosomes to the poles at the first meiotic division in two hybrids and a polyhaploid of Nicotiana. Expected random distributions corrected for the effect of bivalent frequency (SFICAS, 1963)

Hybrid	Plant	Distributions						No. of cells	Mean bivalent frequency	χ^2	Deviation
		12-12	11-13	10-14	9-15	8-16	other				
N. glutinosa × N. sylvestris	E$_1$	59	114	70	32	19	10	304	0.97	15.61*	peaked
	expected	51	94	72	46	25	16				
	E$_2$	58	88	78	38	17	14	293	0.76	6.40$^{n.s.}$	peaked
	expected	49	90	69	45	24	16				
N. glutinosa × N. otophora	D$_7$	68	124	91	39	7	4	333	3.57	9.30$^{n.s.}$	peaked
	expected	65	115	80	45	19	9				
	D$_9$	73	108	76	25	10	5	297	3.23	17.00**	peaked
	expected	57	101	71	40	18	10				
N. tabacum polyhaploid	23 t + 1 pb	56	122	82	70	51	69	450	0.35	83.98**	flattened
	expected	73	136	106	70	38	27				

of the absence of chiasmata. One showed no definite pattern but was nevertheless significantly deviant. It may be a chance deviation. The haploid with a substituted chromosome had a flattened distribution, i.e. most chromosomes tended to move to the same pole.

The clear distinction between these two types of distribution has been observed on more occasions and must be based on some basic segregational mechanism. The movement of an entire set of univalents to one pole (or a group of univalent chromosomes remaining together) can lead to the formation of complete restitution nuclei. When combined with normal behavior of bivalents present in the same cell, it can lead to specialized deviant cytogenetic systems. In the well-known case of cyclic parthenogenetic (thelotokous) aphids, at the end of a series of parthenogenetic reproduction cycles, a transition to sexual reproduction takes place. In addition to sexual females, males must be produced. Whereas the females have two X-chromosomes in addition to twelve autosomes, the males have only one X. In the production of eggs that are to produce the male aphids, one X-chromosome must be eliminated. This is done in a rudimentary form of meiosis: the autosomes remain univalent, but for some reason the X-chromosomes pair and form a bivalent. All univalents collect at one pole but the X-bivalent segregates: one X joins the univalents and the other moves to the other pole where it is included in a minute polar body and disappears.

A similar but even more specialized behavior is exhibited by the plant *Rosa canina*. It is a pentaploid with three identical genomes and two homologous but well differentiated genomes: aaacd $(5x = 35)$ (GUSTAFSSON and HÅKANSSON, 1942; DARLINGTON, 1937, 1965). The three homologous genomes form bivalents exclusively, all other chromosomes remain univalent. There are thus seven bivalents and 21 univalents. In the pollen mother cells most of the univalents are lost and pollen grains with 7, 8, 9 chromosomes, some with higher numbers, are formed. Practically only those with 7 function. In the embryosac mother cell the 21 univalents collect at one pole, and the 7 bivalents segregate normally. 7 chromosomes join the 21 univalents to form a daughter cell with 28 chromosomes. This is the functional cell that forms the embryosac with an egg cell with 28 chromosomes. Fertilization yields a zygote with the original 35 chromosomes.

Different specialized forms of chromosomal elimination and accumulation can occur in several insect species (WHITE, 1973), and preferential segregations as a result of the presence of special, often heterochromatic segments occur in animals and plants. Well known are the accumulation systems of B-chromosomes in numerous animal and plant species and preferential segregation of chromosomes with heterochromatic knobs in maize in the presence of abnormal chromosome 10 (RHOADES, 1952). In principle, such deviant mechanisms may all be considered some form of *meiotic drive*. This term was first used by SANDLER and NOVITSKI (1957) for disturbed segregations of marker genes in chromosome 2 of *Drosophila melanogaster*. When male flies carrying the SD (segregation distorter) factor in this chromosome were testcrossed with females homozygous for recessive markers, segregations of up to 26:1 instead of the expected 1:1 ratio were observed. For a review see PEACOCK and MIKLOS (1973). Such meiotic deviations have usually been studied by genetic methods only, and when accompanied by cytological analysis of meiosis, were mostly of a qualitative character and therefore of limited interest in the present context.

4.2 Orientation and Reorientation in Multivalents; Interchange Heterozygotes and Polysomics

Multivalents originate primarily in two ways: as a result of translocation heterozygosity and as a result of polysomy. A quadrivalent pairing configuration is formed when all homologous segments pair in the heterozygote of a reciprocal translocation, but also in the case of an autotetrasomic. For the orientation of the centromeres to the poles the characteristics of the metaphase configuration are generally of more importance than those of the pairing configurations. In "normal" chiasmate meiosis these characteristics are to a great extent determined by the presence or absence of chiasmata in critical segments (rings, "frying-pans", chains, trivalents etc.). Although translocation quadrivalents and polysomic quadrivalents are very similar in respect to the shapes they can assume, they differ in one important aspect. Whereas in translocation quadrivalents the point of partner exchange is rigorously fixed (perhaps not always at pachytene but certainly in respect to chiasma formation), the point of partner exchange in polysomics is variable. This results in different arrays of metaphase configurations for the two types, but the behavior in respect to orientation is similar for each configuration, independent of its origin from a translocation or from polysomy.

Another important difference is in the *consequence* of orientation. In autotetrasomics, balanced segregation results from any combination of two chromosomes at each pole, in translocation heterozygotes two special chromosomes must be combined to form a balanced set (Fig. 4.2). This means that in translocations two adjacent chromosomes in a quadrivalent may not be combined at the same pole, as this would lead to duplications and deficiencies. Only combinations of chromosomes resulting from "alternate" orientation are balanced. In the autotetrasomic both alternate and adjacent orientations lead to balanced sets of chromosomes. For translocation heterozygotes, two adjacent orientations are distinguished: one in which homologous centromeres coorientate (adjacent I, Fig. 4.2 B), the other in which non-homologous centromeres coorientate (adjacent II, Fig. 4.2 C). Both lead to unbalanced chromosome combinations, but with the breakpoints distally located adjacent II produces the most pronounced unbalance whereas with the breakpoints proximally located, adjacent I may produce the greatest unbalance. Since coorientation is a matter of forces and not of homology, adjacent I and adjacent II tend to occur with the same frequency. There is one notable exception, i.e. when an interstitial chiasma is formed in a quadrivalent of metacentric chromosomes. Then coorientation between homologous centromeres is much more efficient than between non-homologous centromeres, and frequently adjacent II orientation is completely excluded (BURNHAM, 1956). For acrocentric chromosomes this is of less importance. With interstitial

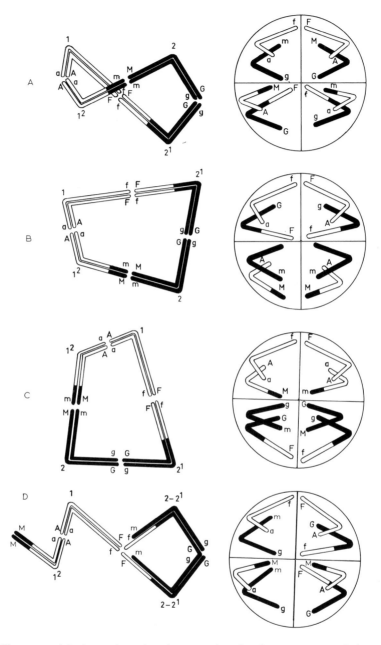

Fig. 4.2A–D. Different quadrivalent orientations in a translocation heterozygote, and the consequences. (A) With alternate segregation of a ring quadrivalent four balanced end products result. (B) With adjacent I orientation, all end products are unbalanced (having a duplication and a deficiency). The same is true for adjacent II orientation (C), although the chromosomal combinations are different. (D) With an interstitial chiasma (cf. Fig. 2.18) two balanced end products (one with the normal, the other with the translocation complement) result, and two unbalanced (duplication-deficiency) ones. No distinction between alternate and adjacent I is possible. Adjacent II is infrequent (SYBENGA, 1972a)

chiasmata, however, the difference between adjacent I and alternate vanishes: both poles receive recombinant chromosomes with a translocation chromatid and a normal chromatid. These separate at anaphase II and result in one balanced and one unbalanced combination (Fig. 4.2 D).

The ratio between the frequencies of alternate and adjacent orientation in ring quadrivalents with random orientation can, in a reasonably realistic way, be derived to be 1:1 both for translocations and tetraploids (SYBENGA, 1972a). In Fig. 4.3 the centromeres are 1, 1^2, 2 and 2^1 respectively. When 1 coorientates with 1^2, 2 and 2^1 have no choice but to coorientate together. There are two possibilities for these two centromeres in relation to 1 and 1^2, one resulting in an alternate ring (Fig. 4.3 A), the other resulting in adjacent I (Fig. 4.3 B). Centromere 1 has another neighbor with which it can coorientate: 2^1. Then 1^2 and 2 coorientate together and again have two possibilities to do so, one resulting in alternate (Fig. 4.3 C), the other in adjacent II orientation (Fig. 4.3 D).

This reasoning is based on a two-by-two orientation and independence of the two pairs of centromeres, which is not realistic for all situations. Two major conditions may disturb the random orientation of the centromeres in a multivalent: the *rigidity* of the multivalent and centromere *reorientation*. Only flaccid multivalents will be able to show frequent alternate orientation. Rigid structures may be the result of stiffness of the chromosomes (SYBENGA, 1968) but also to proximal localization of the chiasmata and of the breakpoint (BURNHAM, 1956). An example of predominant adjacent orientation is that of KHOSHOO and MUKHERJEE (1966) for Canna, where ring quadrivalents tended to orientate adjacent, while chains had a more frequent alternate orientation.

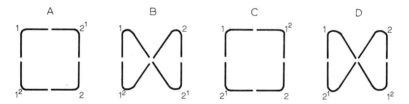

Fig. 4.3 A–D. A reasonably rational way of deriving the relative frequencies of different orientation types with "random" orientation of a translocation ring quadrivalent. Centromeres *1* and *1^2* are homologous, as are *2* and *2^1* (cf. Fig. 4.2). Coorientation involves two adjacent centromeres. Centromere *1* may coorientate with *1^2* (A and B) or with *2^1* (C and D). When *1* coorientates with *1^2*, *2^1* and *2* have two (assumed) independent modes of orientation (A and B) resulting in one adjacent and one alternate ring. When *1* coorientates with *2^1*, centromeres *1^2* and *2* have two independent modes of orientation. This results in a ratio of 1:1 for alternate:adjacent, and also 1:1 for adjacent I:adjacent II. (SYBENGA, 1972a)

Predominant alternate orientation is frequent, and probably mainly caused by an accumulation of alternate types of orientation after centromere re-orientation. The stability of orientation of a centromere on a pole is determined by the pull exerted on it into the opposite direction. The angle between the centromere axis and the spindle axis may play a role in centromere stability. When for some reason the balancing counter-force lapses, the centromere may loose its connection with its pole and gets an opportunity to determine anew to which pole it will orientate. If this happens to be the opposite pole, the centromere *reorientates*. The process of reorientation has been studied in living and in fixed material. Its consequences for the final orientation in multivalents corresponding with anaphase segregation, has been most completely analyzed by ROH-LOFF (1970). This analysis was performed on three reciprocal translocations in *Pales ferruginea* but is in principle applicable to all multivalents (translocations and polysomics) of metacentric chromosomes without interstitial chiasmata, exclusive syntelic orientation, and each centromere orientated to one of the two poles. Under these conditions a ring quadrivalent can have four shapes: alternate ring (tetraeder); flat ring; delta ring (three centromeres to one, the fourth to the other pole); single pole orientation. Now, in a multivalent a centromere can have different "positions", each with its own constant probability of reorientation. The position is determined by the orientation of the two neighboring centromeres: *cis* (*c*) when orientated to the same pole, *trans* (*t*) when orientated to the other (Fig. 4.4). In a ring quadrivalent a centromere can have both

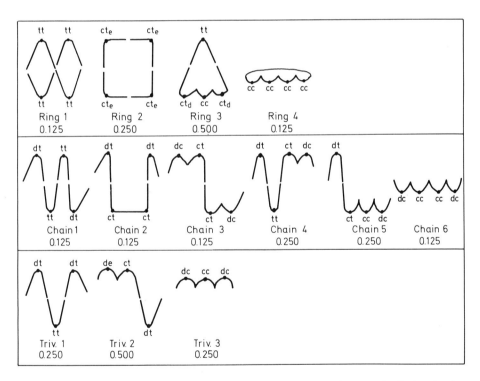

Fig. 4.4. Multivalent shapes (ring quadrivalents; chain quadrivalents and trivalents); the expected frequencies with random (initial) orientation; and for each centromere the position in respect to its two neighbors: *tt*, both neighbors in "trans" position; *ct*, one neighbor "cis" the other trans: ct_e in a flat ring and ct_d in a delta ring; *dt*, one neighbor absent (the centromere is the distal one in a chain), one neigbor trans; *dc*, one neighbor absent (centromere distal) the other cis. (After ROHLOFF, 1970)

its neighbors in the trans-position (tt). This is the position in which all centromeres are in an alternate tetraeder and one centromere in the delta configuration (five together). A centromere can have one neighbor in trans and the other in cis: ct_e for the four centromeres in a flat ring and ct_d for the two outer centromeres in the three parallel centromeres of a delta ring. Finally there is the cc position: one for the central centromere of the three parallel ones in the delta ring and four for the ring with all centromeres to the same pole, a total of five. The sum, of course, is 16 (four centromeres in each of four ring shapes).

There are six chain quadrivalent orientations (Fig. 4.4): (1) alternate; (2) two central centromeres to one pole, two distal ones to the other; (3) two end centromeres to one pole, two to the other; (4) one of the two central centromeres to one pole, three to the other; (5) one end centromere to one pole, three to the other; (6) all centromeres to one pole.

For the chain trivalents there are three orientations: (1) alternate; (2) two adjacent centromeres to one pole, one to the other; (3) three to one pole. The corresponding centromere positions can be readily derived for these types as they have been above for the rings.

The probabilities of reorientation at these positions are real, and (when it is assumed that one reorientation takes place at a time, independently of preceding or following reorientations) will lead to definite probabilities of transition of one multivalent shape into the other in any specified time interval. With a single centromere reorientation, the alternate ring can not be transformed directly into an adjacent ring as this requires two steps, and the transition type is the delta ring. The relative frequencies of the different orientation types move in a complex way towards an equilibrium where the increase of each type is compensated by an equivalent decrease. When multivalents are sampled at specific intervals, the probability of reorientation at each centromere position can be derived from the change in relative frequencies of the different types during the intervals. In ROHLOFF's (1970) study, multivalents were sampled at prometaphase and metaphase, stages which could be determined relatively exactly on the basis of chromosome and equator morphology. Within each stage timing was not well possible. As a third stage the initial (random) orientation was taken, from which the whole process was assumed to start, and where no coorientation has yet taken place. Only considering the rings (Fig. 4.4), it can be seen that there is one possibility for all centromeres to orientate on one pole, one possibility for alternate orientation, two for adjacent orientation and four for delta orientation. This results in initial frequencies of 0.125 for alternate; 0.250 for adjacent; 0.500 for delta and 0.125 for all-to-one-pole. These theoretical frequencies for the four ring types, together with the prometaphase and metaphase observations are given in Table 4.6 and graphically in Fig. 4.5A, for the three translocations. For the chains and trivalents ROHLOFF gives similar data.

The stochastic processes of reorientation leading to these frequencies can be simulated by iteration, estimating the approximate reorientation frequencies of the centromeres in their positions in each multivalent shape. Such a simulation does not prove that reorientation in spermatocytes of *P. ferruginea* actually takes place as described, but it shows that by reorientation with the proposed probabilities the observed multivalent frequencies can be realized.

For ring quadrivalents the Markow chain describing the reorientation process has four possible situations corresponding with (1) the alternate tetraeder; (2) the adjacent ring; (3) the delta ring and (4) the ring with all centromeres towards one pole. When P_{ij} is the probability that a ring has orientation type j at time $n+1$ under the condition that at time n it was of type i, than the matrix of transition probabilities is

$$
\begin{matrix}
P_{11} & P_{21} & P_{31} & P_{41} \\
P_{12} & P_{22} & P_{32} & P_{42} \\
P_{13} & P_{23} & P_{33} & P_{43} \\
P_{14} & P_{24} & P_{34} & P_{44}
\end{matrix}
$$

For $P_{n+1} = A \cdot P_n$, the transformation equation describing the probabilities of change of one ring shape into another is that of Table 4.7. For chains and trivalents the corresponding equations can be constructed. Now, in the matrix given above, P_{11} indicates that the tetraeder remains a tetraeder; P_{12} indicates

Table 4.6. The theoretical initial and the observed prometaphase and metaphase frequencies of the orientation types of Fig. 4.4 for three reciprocal translocations in *Pales ferruginea* (ROHLOFF, 1970)

Orientation type (Fig. 4.4)	Theoretical initial frequency	Translocation 177 promet.	metaph.	Translocation 181 promet.	metaph.	Translocation 215 promet.	metaph.
Rings							
1	0.125	0.420	0.665	0.310	0.485	0.565	0.785
2	0.250	0.479	0.320	0.586	0.475	0.334	0.205
3	0.500	0.101	0.015	0.103	0.004	0.101	0.010
4	0.125	0.000	0.000	0.000	0.000	0.000	0.000
Total rings		584	275	435	198	701	297
Chains							
1	0.125	0.446	0.695	0.425	0.594	0.408	0.620
2	0.125	0.308	0.221	0.287	0.252	0.225	0.279
3	0.125	0.073	0.031	0.112	0.077	0.157	0.052
4	0.250	0.152	0.053	0.131	0.067	0.157	0.043
5	0.250	0.021	0.000	0.045	0.010	0.052	0.007
6	0.125	0.000	0.000	0.000	0.000	0.000	0.000
Total chains		565	226	771	298	853	305
Trivalents							
1	0.25	0.658	0.761	0.568	0.730	0.600	0.909
2	0.50	0.342	0.239	0.432	0.269	0.400	0.091
3	0.25	0.000	0.000	0.000	0.000	0.000	0.000
Total trivalents		158	71	259	89	50	11
Total multivalents		1,307	572	1,465	585	1,604	613

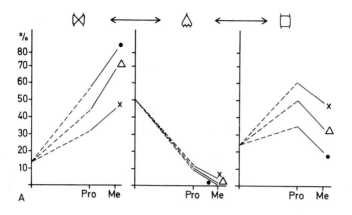

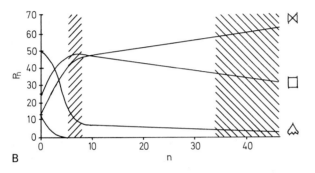

Fig. 4.5.(A) The change in frequency of three orientation types of a ring quadrivalent, from the assumed random initial orientation *via* promethaphase to the (final) metaphase orientation. Three different translocations: 177△; 181 × ; 215● in *Pales ferruginea*. (B) Simulation curves for the frequency changes of the ring orientations. Relative frequencies on the ordinate, number of reorientation steps (not necessarily of equal duration) on the abscissa. The initial situation at the origin; the shaded areas indicate period of observation at prometaphase and metaphase respectively. Translocation 177 in *Pales ferruginea*. (After ROHLOFF, 1970)

Table 4.7. Transformation equation describing the probabilities of change of one ring shape into another as a consequence of centromere reorientation. The four ring types are numbered I-IV; $n+1$ is the situation one reorientation step after n (ROHLOFF, 1970)

$$
\begin{bmatrix}
p_{n+1}^{I} \\
p_{n+1}^{II} \\
p_{n+1}^{III} \\
\\
\\
p_{n+1}^{IV}
\end{bmatrix}
=
\begin{bmatrix}
1-4\bar{p}_{(tt)} & 0 & \bar{p}_{(cc)} & 0 \\
0 & 1-4\bar{p}_{(ct_o)} & 2\bar{p}_{(ct_d)} & 0 \\
4\bar{p}_{(tt)} & 4\bar{p}_{(ct_o)} & 1-\bar{p}_{(tt)} & 4\bar{p}_{(cc)} \\
& & -2\bar{p}_{(ct_d)} & \\
& & -\bar{p}_{(cc)} & \\
0 & 0 & \bar{p}_{(tt)} & 1-4\bar{p}_{(cc)}
\end{bmatrix}
\begin{bmatrix}
p_n^{I} \\
p_n^{II} \\
p_n^{III} \\
\\
\\
p_n^{IV}
\end{bmatrix}
$$

that a tetraeder changes into an adjacent ring, which is impossible in one step, and therefore $\bar{p}=0$; P_{13} is the transition (by one reorientation) of an alternate tetraeder into a delta ring with a probability equal to $4\bar{p}_{tt}$, etc., cf. Table 4.7. Now the interval is chosen such that the sum of the 16 possible $\bar{p}$ values (four centromeres in each of the four configuration shapes) equals unity, each configuration shape counting equal. Thus: $5 \times tt + 4 \times ct_e + 2 \times ct_d + 5 \times cc = 1$ for the ring. By iteration (starting from the initial knowledge that $cc > ct_d > ct_e > tt$) values for these parameters were computed that best fitted the initial, the prometaphase and the metaphase observations. These values are given in Table 4.8,

Table 4.8. Reorientation probabilities of the centromeres in the four different positions in the ring quadrivalents, found by iteration, using the equations of Table 4.7 on the basis of the theoretical initial and observed prometaphase and metaphase values of Table 4.6 (ROHLOFF, 1970). The reorientation probabilities add to 1 since there are $5tt$ positions, $4ct_e$ positions, $2ct_d$ positions and $4cc$ positions

Centromere position	tt	ct_e	ct_d	cc
Translocation				
215	0.0001	0.005	0.025	0.186
177	0.001	0.006	0.072	0.165
181	0.001	0.009	0.135	0.138

and the curves showing the suggested changes in configuration shape frequencies in Fig. 4.5B. It may help in interpreting these curves when it is seen that in the first interval, for instance for translocation 181, the alternate tetraeder can only be formed from the delta ring by a cc reorientation. The probability of this reorientation is 0.186 (Table 4.5), and the frequency of the delta ring is initially 0.5. Thus the alternate tetraeder increases by $0.186 \times 0.5 = 0.093$. At the same time it decreases by $4 \times tt = 4 \times 0.001 = 0.004$. The net increase is 0.089. Added to the initial 0.125, the frequency at the end of the first interval is $0.125 + 0.089 = 0.214$.

It should be noted that there is not necessarily a reorientation at each interval. Especially at metaphase where the shapes with many tt centromeres tend to predominate, the total stability is high and change is slow. For instance, at prometaphase of translocation 177 (Tables 4.6; 4.8), the tt centromeres with a probability of reorientation of 0.001 occur with a frequency of 0.420×4 (tetraeder) $+ 0.101 \times 1$ (delta) which contributes to reorientation with a frequency of 0.00178. For ct_e the reorientation frequency is 0.479×4 (adjacent) $\times 0.006 = 0.01150$. From ct_d centromeres the contribution to reorientation is 0.101×2 (delta) $\times 0.072 = 0.01454$ and from cc centromeres it is 0.101×1 (delta) $\times 0.165 = 0.01667$. At this stage the total probability of reorientation per interval is only 0.04449. At metaphase the equilibrium is approached at which $P_n = A \cdot P_n$, the limit of the equation of Table 4.7. It need not necessarily be reached, since it is very well possible that anaphase starts earlier.

This is at present the most complete model for the analysis of centromere reorientation, using configuration frequencies at different stages, and for this reason has received considerable attention here. A comparably sophisticated mathematical model of chromosome segregation from multivalents has been worked out by DOUGLAS (1968b, 1968c). One of its bases is the hypothesis of presegregation, i.e. the pole towards which a centromere ultimately migrates is determined before actual movement sets in. Any reorientation merely serves to correct an initial error of orientation. The model has been constructed for explaining observed segregation ratios and is not immediately suitable for analysis of meiotic configurations.

ROHLOFF'S (1970) model is not perfect, since some of the assumptions on which it is based are not always acceptable. One is the constancy in time of the stability of the different centromeres in relation to each other. There are indications that the stability of a centromere increases with the time it is orientated on a specific pole. Then the centromere positions (*tt*, for instance) with the greatest stability will attain an even greater relative stability during metaphase. A second assumption which may be acceptable for *P. ferruginea* but not for several other organisms is the absence of linear orientation, in which one chromosome is tied up between two others and fails to orientate at all, perhaps due to the stretching of its centromere. The active axis of this centromere is perpendicular to the spindle axis which inhibits a polar orientation. Since in a linear quadrivalent there are also two *tt* centromeres, the stability may be considerable (Fig. 4.6). This orientation may lead to either of the two adjacent segregations or to numerical nondisjunction, depending on where the non-oriented centromeres go. It is probable that linear orientations are possible only where the spindle is long relative to the length of the chromosomes involved.

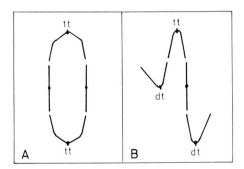

Fig. 4.6A and B. Linear arrangement of chromosomes in ring (A) and chain (B) quadrivalents

A third assumption, independence of centromeres in respect to reorientation, is open to more fundamental criticism. It is in fact not unreasonable to assume that if one centromere reorientates, another specific centromere is induced to do the same. This is the reason that SYBENGA (1968, 1972a) did not consider the (very) infrequent delta configuration to be a true orientation type, but merely a transition phase between the alternate and adjacent orientations. If orientation stability is determined by a pull in the opposite direction, it would be expected that lapse of orientation of one centromere in an adjacent ring (for instance because it happens to approach the pole too closely) will not only give this centromere the opportunity to reorientate, but will also immediately disturb the orientation of the opposite centromere. This now gets the opportunity to change its orientation and both in conjunction can choose their respective (new) orientation. This new orientation may be such that an alternate orientation results, or the adjacent ring may be reestablished. The other two centromeres coorient together and are not affected. In an alternate ring (tetraeder), when the orientation of one centromere lapses, the opposite centromere still has an orientation-partner and is not induced to reorientate. It may even cause the originally hesitating centromere to resume its orientation. Only when accidentally two centromeres are disturbed simultaneously, will reorientation be expected. When simultaneous instability is a condition for reorientation, it depends on the duration of the instability, unless the cause of the instability systematically acts on two centromeres at the same time, for instance when the entire configuration comes too close to one pole.

These considerations make it understandable that chains are more liable to change their shape than rings, and that the equilibrium position is attained later for chains than for rings. Then anaphase may find more rings in alternate orientation than chains. This was indeed found by SYBENGA (1968) (Table 4.9), and the data of GAIRDNER and DARLINGTON (1931) shown in the same table, demonstrate the same trend. In other cases, however, chains are found to orientate alternately more frequently than rings, and this may be due to a greater stiffness of rings, where stiffness plays a role in reorientation.

The assumption of coordinated reorientation suggests a similar high stability of *tt* centromere positions, as observed (but not *predicted*) by ROHLOFF (1970). This approach requires some significant alterations in ROHLOFF'S model, but these have not yet been worked out.

Among the translocations reported by ROHLOFF (1970) one (215) clearly had a higher frequency of alternate orientation than the others. It was also the most symmetric one in respect to chromosome arm length. Clearly factors such as length of the interchanged segments and of the interstitial segments, relative chromosome arm length, chiasma position (BURNHAM, 1956), flexibility of the chromosome, and time available for reorientation are important for the final result. It is not amazing, therefore, that a great variation in orientation (and consequently of segregation) is observed, and that the genetic component in this variation is significant. In autotetraploids selection for fertility tends to increase alternate orientation, which gives a better stability and a reduced chance of 3:1 segregation. It is probable that not the alternate orientation as such is so much better, but that the conditions leading to an accumulation of alternately orientated configurations also reduce linear and indifferent orientations resulting in a 3:1 segregation. In established natural autopolyploids alternate orientation can also be considerably more frequent than in their newly induced parallels. McCOLLUM (1958), for instance found in Dactylis for induced tetraploids in four species on an average among 2,083 quadrivalents 850 (40.8%) with alternate orientation. In three hybrids between these the percentage alternate among 1,155 quadrivalents was 40.1. In three natural autopolyploids among 834 quadrivalents the percentage alternate was significantly higher: 47.8%, and in a hybrid between two of these 62.7% (826 quadrivalents). A hybrid between an induced and a natural autotetraploid had 47.5% alternate among 339 quadrivalents.

In translocation heterozygotes too, the frequency of alternate orientation varies greatly between species and is also under genetic control within species. Selection for fertility leads to increased alternate orientation (LAWRENCE, 1958) in rye. SUN and REES (1967) reported on an experiment of direct selection on disjunction frequency. Heterozygotes of two interchanges (A and B) and their hybrid (AB) were used and in F_2 and F_3 the plants with highest and lowest alternate orientation were selected to produce the next generation. There was a constant increase in alternate orientation in *both* high and low lines, which shows that the response is not without complications (Table 4.10). The cause of this effect apparently is a general decrease in chiasma frequency accompanying inbreeding, and favoring alternate disjunction. A correlation between high frequency of alternate orientation and low chiasma frequency had earlier been

Table 4.9 (A). Reorientation of ring and chain quadrivalents in two translocation heterozygotes (248 and 273) of rye during metaphase I. Early and late cells from different anthers of same inflorescence; a "late" anther has a high frequency of cells at stages later than metaphase I (mainly anaphase I) (SYBENGA, 1968)

Trans-location	M I	Rings alternate	Rings adjacent	Chains alternate	Chains adjacent	Total	Contingency $\chi_1^2 = 59.61$
248	Early	143	45	119	116	413	
	Late	155	11	193	49	409	
	Total	298	56	312	155	821	

Contingency χ_1^2 alternate-adjacent/early-late = 51.34
Contingency χ_1^2 rings-chains/early-late = 1.96 (not sign.)
Interaction $\chi_1^2 = 6.312$

273	Early	727	117	100	43	987	Contingency $\chi_1^2 = 89.48$
	Late	311	28	141	13	493	
	Total	1,038	145	241	56	1,480	

Contingency χ_1^2 alternate-adjacent/early-late = 17.46
Contingency χ_1^2 rings-chains/early-late = 57.50
Interaction $\chi_1^2 = 14.52$

Table 4.9 (B). Orientation of rings and chains in a translocation heterozygote of *Campanula persicifolia* (GAIRDNER and DARLINGTON, 1931)

	Rings	Chains	Total	
Alternate	173	14	187	
Adjacent	106	26	132	$\chi_1^2 = 10.52$
Total	279	40	319	

Table 4.10(A). The mean disjunction frequencies (after angular transformation) of selected F_2 plants and their F_3 and F_4 progenies (5–6 plants for each; 20 PMC per plant) of rye, heterozygous for reciprocal translocations A and B and their combination (SUN and REES, 1967). Selection was on high and low alternate disjunction frequencies. (B) The mean chiasma frequencies for the same material. (C) The relationship between the (alternate) disjunction frequency and number of chiasmata in the translocation quadrivalent in the same material (REES and SUN, 1965) in F_2 and F_3

(A)

Trans-location	F_2 level of disjunction	F_2	F_3	F_4
A	high	60.0	63.4	79.56
	low	45.0	59.1	77.48
B	high	67.2	71.9	69.40
	low	47.9	60.4	71.60
A + B	high	77.1	71.6	97.26
	low	49.3	64.4	67.70

(B)

F_2 level of disjunction	Trans-location	F_3	F_4	Mean
High	A	12.90	12.70	12.05
	B	12.98	12.72	
	A + B	12.53	12.48	
Low	A	13.25	12.85	13.06
	B	13.30	12.95	
	A + B	13.04	12.99	

(C)

Transloc.	No. of chiasm.	Mean disjunction (angular)	Transloc.	No. of chiasm.	Mean disjunction (angular)
A	3	65.9	A	3	75.0
	4	58.0		4	61.9
	5	—		5	45.0
B	3	68.2	B	3	72.5
	4	61.1		4	66.8
	5	40.9		5	31.1
A + B	3	70.3	A + B	3	73.0
	4	68.7		4	65.7
	5	45.0		5	32.3
	F_2			F_3	

shown to exist in rye translocation heterozygotes. SUN and REES (1967) conclude that "the selection was effective in establishing distinct lines with *relatively* high and low disjunction frequencies. At the same time the inbreeding practised during the selection gave rise to a situation that may best be described in terms of genetic inertia (MATHER, 1953), where variation in one character, disjunction, is restricted and in this case largely determined by the release of variation in the other, the chiasma frequency". Up to a limit decrease in chiasma frequency will then favor fertility but beyond this limit, as in polyploids, low chiasma frequency will cause univalent formation and thus ultimately cause fertility to decrease. In interchange heterozygotes lack of chiasmata will also cause the formation of bivalents which as a result of absence of coordinated coorientation will not be able to produce more than 50% alternate segregation. The relation between low chiasma frequencies and high alternate orientation in rye is not simple: SYBENGA (1968) has shown that chains (with fewer chiasmata) tend to have a lower frequency of alternate orientation than rings. It is well possible that within rings, a lower chiasma frequency results in fewer interstitial chiasmata and in general not more than one chiasma in an arm, which improves the flexibility of the ring. This aspect apparently has not been explicitly considered in the analysis of SUN and REES (1967) and of REES and SUN (1965), but from their data (Table 4.10) it appears that the greatest effect results from the transition of five to four chiasmata per quadrivalent. This suggests that proximal chiasmata are important in maintaining a rigid structure. As a matter of fact, with five chiasmata the orientation is predominantly *not* alternate.

The behavior of multivalents of acrocentric chromosomes is necessarily different from those of metacentric chromosomes since interstitial chiasmata are a prerequisite for the formation of multivalents (Fig. 4.7). Few data on metaphase

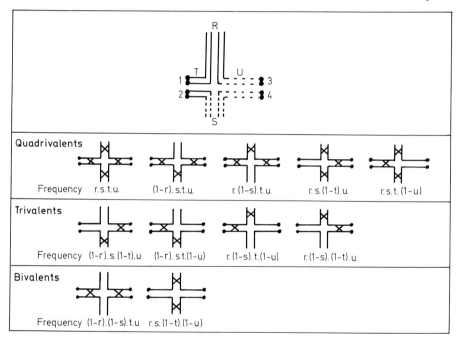

Fig. 4.7. Quadrivalent pairing and various chiasma positions in a heterozygote for a reciprocal translocation between two acrocentric chromosomes. Only segments R, S, T and U are available with r, s, t and u as frequencies of being bound (cf. Figs. 2.18; 2.19). Interstitial chiasmata together with chiasmata in the translocated segment are a prerequisite for a multivalent to be maintained into metaphase I. The frequencies of the different configuration types are given in terms of r, s, t and u (cf. Fig. 4.8)

orientation have been reported. In some animals, particularly the thoroughly analyzed house mouse, *Mus musculus*, although many acrocentric multivalents have been studied, with the techniques necessary to demonstrate any detail of the chromosomes, the original orientation is usually destroyed. The example given here, therefore, again is of a plant (rye, *Secale cereale*) where the acrocentric B chromosomes only infrequently have a chiasma in the very short arm, and where multivalent formation depends primarily on partner exchange and formation of simultaneous interstitial and terminal chiasmata in the long arm. In tetrasomics for these B-chromosome, KISHIKAWA (1965, 1966) reported on the orientation of the centromeres in quadrivalents. In several different plants (one of which triploid), the distribution of Table 4.11 was found (cf. Table 2.15 and Fig. 2.10). With completely random orientation, in $\frac{1}{8}$ of the cells all four centromeres are expected to orientate to the same pole. In $\frac{4}{8}$ three go to one and one to the other pole and in $\frac{3}{8}$ two to one, and two to the other pole. No case of orientation of all centromeres to one pole was observed at metaphase, which is not unexpected if orientation always requires the interaction between two centromeres, but also the ratio between 2-2 and 3-1 was far different from expected. The great excess of 2-2 orientation suggests considerable reorientation. Even though with 3-1 orientation all centromeres would be expected to have a stable position, apparently when three centromeres move to one pole, they come too close to it too early, and lose their stability. As soon as one centromere reorientates, the stability is increased, as pair-wise reorientation is not necessary. An accumulation of 2-2 orientation is thus accomplished.

Table 4.11. The frequencies of the three types of metaphase I orientation of sub-acrocentric B-chromosomes (tetrasomic) in diploids and triploids of rye (KISHIKAWA, 1965, 1966) (infrequent short arm associations neglected)

Orientation	$3n+4$B	$2n+4$B	$2n+4$B	$2n+4$B	Total	Expected (3:4:1)
⅄	103	548	174	189	1,014	410.25
Ψ	6	44	11	19	80	547.00
⩘	0	0	0	0	0	136.75
Total number	109	592	185	208	1,094	1,094.00
of cells	200	800	571	450	2,021	
analyzed[a]	(1 plant)	(4 plants)	(1 plant)	(1 plant)	(7 plants)	

Heterogeneity $\chi^2_3 = 0.340$.

[a] Not in all cells the B chromosomes were associated in configurations of four chromosomes.

In the 2-2 orientation of a tetrasomic multivalent (acrocentric or not), anaphase disjunction will always lead to balanced spores. In a translocation multivalent, however, the specific combination of two chromosomes out of the four present is of great consequence for the genetic composition of the resulting spores. When the different chromosomes can be recognized, the type of segregation can be analyzed at anaphase I or metaphase II, but since in acrocentrics there is always at least one interstitial chiasma this analysis is not without complications. Because the MI configuration is symmetric, any two of the four chromosomes may combine with approximately equal frequency, unlike with metacentric chromosomes. Thus there are three possible combinations of two chromosomes which are formed with theoretically equal probability. In one, homologous centromeres segregate (Fig. 4.7, 4.8), in two, non-homologous centromeres. What this results in, depends on the segments in which chiasmata have occurred. To keep the four chromosomes together in one quadrivalent, there must be chiasmata in at least three segments of the four. Out of four segments, four combinations of three can be formed and there is a fifth with chiasmata in all four segments (Fig. 4.7). With chiasmata in R, S, T and U; in R, T and U; and in S, T and U the effect on anaphase I (and M II) chromosome constitution is identical, and these are taken together in one category when A I or M II segregations are considered. The R, S, T and R, S, U chiasma combinations produce different A I and M II chromosome shapes, as can be seen in the figure, and therefore represent separate recognizable categories. The relative frequencies of these categories depend entirely on the system of chiasma formation, which can be analyzed at diakinesis. In order to have exact information on the location of the chiasmata, the centromeric ends of the chromosomes must be recognizable. At prometaphase and metaphase this is possible on the basis of centromere stretching, but at diakinesis there are difficulties. Then the centromeres may be marked by Giemsa c-banding, as is readily done in *Mus musculus* (POLANI, 1972; DE BOER and GROEN, 1974).

At anaphase I and metaphase II (because of the interstitial chiasmata) no distinction can be made between adjacent I and alternate, but adjacent II gives a recognizably different chromosome combination. Different chromosome shapes result from the three different chiasma combinations, and when at least some of the chromosomes can be distinguished there can be no mistake as to which combination of chromosomes is a result of adjacent II segregation. Fig. 4.8 gives the segregation types for quadrivalents, and also for trivalents, bivalents and univalent combinations on the assumption of 2-2 segregation. In the cases with a chiasma in both interstitial segments, the two chromatids of all four A I (or M II) chromosomes are different. If there is a chiasma in only one, as in two types of (chain) quadrivalent and in trivalents, there is one interkinesis nucleus with different chromatids and one with identical chromatids. Since translocations in which the chromosomes can be classified at A I and M II (even if it is only one) must have a considerable difference between chromosomes, it is usually also possible to distinguish the frequency with which particular types of chains and trivalents are formed at diakinesis or metaphase I and to use the information thus obtained on chiasmata in interpreting A I or M II segregations. With two chiasmata in an interstitial segment, the equivalent of one chiasma results with the two disparate types and the equivalent of no chiasma with the reciprocal and complementary types. This results in a net decrease of the frequency of unequal chromatids in a chromosome. When the frequency of double chiasmata can be estimated they can be corrected for, but the assumption of no chromatid interference may not always be correct (cf. Section 2.1.1.2). When no estimate is possible, it is useful at least to be aware of the effect. Double chiasmata in interstitial segments tend to be very infrequent, however.

When the four chromosomes of the translocation complex can be recognized, it is relatively simple to estimate the frequency of adjacent II on the basis of anaphase I—metaphase II segregations, using Fig. 4.8 and adequate metaphase

I information. Loss of univalents from trivalent-univalent and other univalent combinations and 3-1 segregations must be given separate treatment.

Often, only one of the translocation chromosomes is recognizable and then the analysis is less complete and statistically less efficient. The possibility of estimating adjacent II frequencies remains. Say that in Fig. 4.8, only the shortest chromosome can be recognized, then it can occur at A I (or M II) as two identical chromatids or, if there has been a chiasma in segment T, in combination with a long chromatid. The only recognizable combination which can only be explained by adjacent II, and cannot equally well be due to alternate/adjacent I orientation is the combination of two chromosomes, each with one short and one long chromatid, assuming separated M II cells as in most animals. When the two M II cells of the same meiocyte are combined as in most plants, a somewhat better estimate can be obtained but the principle remains the same. The total frequency of this type, with m as the frequency of adjacent II orientation is

$$1/2m\{r \cdot s \cdot t \cdot u + r \cdot (1-s) \cdot t \cdot u + (1-r) \cdot s \cdot t \cdot u + r \cdot s \cdot t \cdot (1-u)$$
$$+ r \cdot (1-s) \cdot t \cdot (1-u) + (1-r) \cdot s \cdot t \cdot (1-u)\} = 1/2mt(r+s-rs). \quad (4.1)$$

Here, r and s are the frequencies of the translocated segments R and S having a chiasma and t is the frequency of the interstitial segment involving the marker chromosome having a chiasma. This equation applies to quadrivalents which were rings at diakinesis, and those chains in which a chiasma failed in one of the translocated segments. Conversion of Eq. (4.1) to $1/2mt\{1-(1-r)(1-s)\}$ demonstrates that this A I (or M II) configuration can appear only when there is a chiasma in segment T, and not when both R and S fail to have chiasmata.

An estimate of the frequency of adjacent II segregation based on M II classifications in mouse translocation heterozygotes is given by DE BOER (1975).

For metacentric chromosomes this approach is usually not applicable, since it is based on the occurrence of interstitial chiasmata which practically exclude adjacent II (at the same time making alternate and adjacent I equivalent), when the non-translocation arms also have chiasmata. It remains possible of course, if different chromosomes of the translocation complex can be recognized, to estimate the frequencies of the A I segregation types from M II classifications. This has been done for the onion fly (*Hylemya antiqua*) by VAN HEEMERT (1974).

In translocations involving the nucleolar chromosome, the nucleolus can be used as a marker of segregation (Fig. 4.9). When the nucleolus is not located in the translocated segment, i.e. in an arm not involved in the translocation or in an interstitial segment, alternate and adjacent I will result in normal segregation of the nucleolus. Then both daughter nuclei at interkinesis or all four end products at the tetrad stage will have a nucleolus. With adjacent II, however, one of the two interkinesis nuclei and two of the tetrad nuclei will have two nucleoli, the other none. This can usually readily be scored. When, on the other hand, the nucleolus is located in the translocated segment, alternate and adjacent II give "normal" nucleolus segregation, whereas adjacent I results in the absence of nucleoli in half of the cells and double the number in the remainder. An example is given in Table 4.12, for maize (BURNHAM,

1950, 1962). Wu and Pi (1968) reported on the distribution of nucleoli in pollen tetrads in heterozygotes for five translocations involving the nucleolar chromosome in Sorghum, without specifying the segment in which the nucleolus was situated. Among 599 tetrads, 458 were normal and 141 had two nuclei devoid of nucleoli. The percentage deviant types thus was 25.54, resulting either from adjacent I or from adjacent II orientation, depending on the location of the nucleolus. The observed pollen abortion was exactly twice this frequency: 51.20%. Apparently, half the multivalents had orientated alternately, the other half adjacently. Of the latter, half again adjacent I, the other half adjacent II. Since adjacent I and adjacent II were of equal frequency, it may be concluded that no chiasmata were formed in the interstitial segments. This is confirmed by

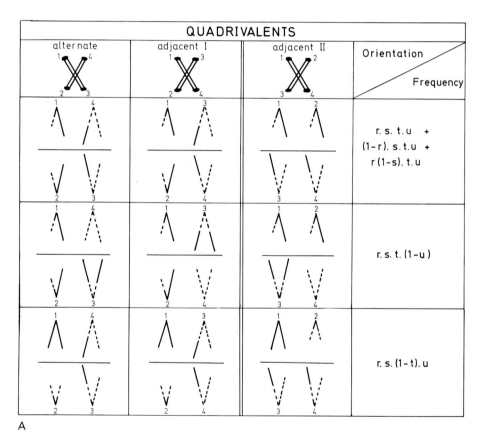

A

Fig. 4.8 A–C. Metaphase I orientation and resulting anaphase I segregation for quadrivalents (A), trivalents with univalent (B) and bivalent pairs (C) in a heterozygote for a reciprocal translocation between acrocentric chromosomes. The frequencies given are those of Fig. 4.7 and are based on the frequency of chiasmata in the four segments, assuming a maximum of one in the interstitial segments. When the translocated chromosomes are distinctly different from the original chromosomes, the resulting A I and even A II segregations can be used to detect adjacent II orientation, and to estimate its frequency when information on chiasma frequency is available. (Cf. Eq. (4.1))

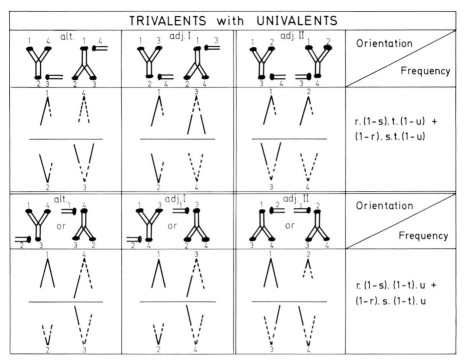

TRIVALENTS with UNIVALENTS			
alt.	adj. I	adj. II	Orientation / Frequency
			$r.(1-s).t.(1-u) + (1-r).s.t.(1-u)$
alt. or	adj. I or	adj. II or	Orientation / Frequency
			$r.(1-s).(1-t).u + (1-r).s.(1-t).u$

B

BIVALENTS			
alt. or	adj. I	adj. II	Orientation / Frequency
a			$r.s.(1-t).(1-u) +$ $\big(r.(1-s).(1-t).(1-u). +$ $(1-r).s.(1-t).(1-u) +$ $(1-r).(1-s).(1-t).(1-u)$ $=$ univalents$\big)$
b			$(1-r).(1-s).t.u.$

C

Table 4.12A and B. The distribution of nucleoli in pollen tetrads of translocation heterozygotes of maize and the derived frequencies of the orientation types of the metaphase I multivalents. (BURNHAM 1950, 1962)

(A) Nucleolus in the non-translocation arm (chromosome 6) (cf. Fig. 4.9A). Tetrads having one spore without nucleolus result from unequal segregation. For other columns free spores and tetrads scored together

Chrom.	Interstitial segment	Total spores	Spores without nucleolus	% tetrads 1 spore no nucleolus	% adj. II	Pollen Total	% aborted
2 and 6	long	4,059	197	1.1	9.8	2,628	47.0
2 and 6	long	6,590	10	0.0	0.3	3,383	48.1
1 and 6	short	16,104	1,251	0.1	15.5	10,291	51.5
5 and 6	short	2,197	278	0.5	25.4	3,094	50.2

(B) Nucleolus in the translocated segment (chromosome 6) (cf. Fig. 4.9B, C). With high frequency of crossing-over in interstitial segments no proper estimate of adj. I and alternate frequencies possible

Chrom.	Interstitial segments	Total tetrads	Tetrads with nucleoli in: 4 cells	3 cells	2 cells	% cross-over interstit. segment	Pollen abortion obs.	exp.	% orientation in adj. 2	no cross-over tetrads adj. 1	alt.
5 and 6	long, short	4,229	809	2,678	742	63.3	48.1	49.3	0.0	—	—
4 and 6	long, short	1,122	246	697	179	62.1	49.8	47.1	2.7	—	—
2 and 6	short, short	2,361	1,738	25	598	1.1	51.6	25.9	26.0	25.3	48.7
6 and 10	short, short	2,374	1,819	126	429	5.3	42.9	20.7	23.4	19.1	57.5

the fact that no tetrads with a single nucleus lacking a nucleolus were reported. An alternative explanation for the latter can of course be that the nucleolus in all translocations was situated in the non-translocated arm.

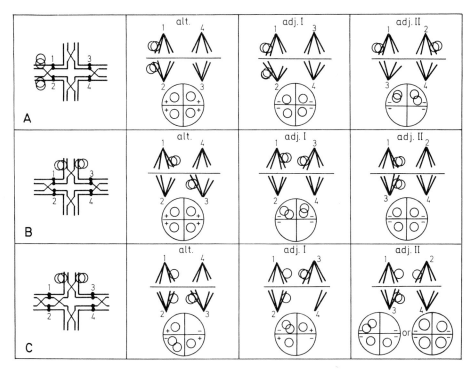

Fig. 4.9 A–C. Nucleolar segregation in spore tetrads. (A) Nucleolus in non-translocation arm of metacentric chromosome. Alternate orientation results in four balanced spores (+) each with a nucleolus; adjacent I in four unbalanced spores (−) each with a nucleolus; adjacent II in four unbalanced spores (−), two of which have 2 nucleoli and two none. (B) Nucleolus in translocated segment, no interstitial chiasmata. Alternate: four balanced spores (+) each with nucleolus; adjacent I: four unbalanced spores (−), two with 2 nucleoli, two with none; adjacent II: four unbalanced spores (−), each with a nucleolus. (C) Nucleolus in translocated segment, one interstitial chiasma. Alternate (=adjacent I): two balanced spores each with one nucleolus, one unbalanced spore with two nucleoli and one unbalanced spore with none; adjacent II: four unbalanced spores, two with one nucleolus, one with two and one with none; or four unbalanced spores, each with one nucleolus

References

BANTOCK, C. R.: Localization of chiasmata in *Cepea nemoralis*. Heredity **29**, 213–221 (1972).

BASAK, S. L., JAIN, H. K.: Autonomous and interrelated formation of chiasmata in Delphinium chromosomes. Chromosoma **13**, 577–587 (1963).

BASAK, S. L., JAIN, H. K.: The interchromosome distribution of chiasmata in interchange heterozygotes of Delphinium. Heredity **19**, 53–61 (1964).

BAUER, H., DIETZ, R., RÖBBELEN, C.: Die Spermatozytenteilungen der Tipuliden. III Mitteilung. Das Bewegungsverhalten der Chromosomen in Translocationsheterozygoten von *Tipula oleracea*. Chromosoma **12**, 116–189 (1961).

BAYLISS, M. W., RILEY, R.: An analysis of temperature-dependent asynapsis in *Triticum aestivum*. Genet. Res. **20**, 193–200 (1972a).

BAYLISS, M. W., RILEY, R.: Evidence of premeiotic control of chromosome pairing in *Triticum aestivum*. Genet. Res. **20**, 201–212 (1972b).

BELLING, J., BLAKESLEE, A. F.: The configurations and sizes of the chromosomes in trivalents of 25-chromosome Daturas. Proc. Nat. Acad. Sci. Wash. **10**, 116–120 (1924).

BELLING, J., BLAKESLEE, A. F.: The assortment of chromosomes in haploid Datura. La Cellule **37**, 353–365 (1927).

BENDER, K., GAUL, H.: Zur Frage der Diploidisierung autotetraploider Gerste. Z. Pflanzenzüchtung **56**, 164–183 (1966).

BENNETT, E. S.: The origin and behavior of chiasmata. XIV. *Fritillaria chitralensis*. Cytologia **8**, 443–451 (1938).

BENNETT, M. D., REES, H.: Induced variation in chiasma frequency in rye in response to phosphate treatments. Genet. Res. **16**, 325–331 (1970).

BODMER, W. F., PARSONS, P. A.: Linkage and recombination in evolution. Adv. Genet. **11**, 1–100 (1962).

BOER, P. DE: Male meiotic behavior and litter size of the T(2;8)26H and T(1;13)70H mouse reciprocal translocations. In press (1975).

BOER, P. DE, GROEN, A.: Fertility and meiotic behavior of male T70H tertiary trisomics of the mouse *(Mus musculus)*. Cytogenet. Cell Genet. **13**, 489–510 (1974).

BRANDHAM, P. E.: Inversion heterozygosity and sub-chromatid exchange in *Agava stricta*. Chromosoma **26**, 270–286 (1969).

BRANDHAM, P. E.: The consequences of crossing-over in pericentric inversions in acrocentric chromosomes. Heredity **25**, 125–129 (1970).

BROWN, M. S.: The implications of pachytene and metaphase pairing for species differentiation. Proc. 10 Int. Congr. Genet. **2**, 36–37 (1958).

BROWN, S. W., ZOHARY, D.: The relationship of chiasmata and crossing over in *Lilium formosanum*. Genetics **40**, 850–873 (1955).

BROWN, W. V.: Textbook of Cytogenetics. Saint Louis: The C. V. Mosby Company 1972.

BROWN, W. V., STACK, S. M.: Somatic pairing as a regular preliminary to meiosis. Bull. Torrey Bot. Cl. **95**, 369–378 (1968).

BURNHAM, C. R.: Chromosomal interchanges in maize: reduction of crossing-over and the association of non-homologous parts. Amer. Nat. **68**, 81–82 (1934).

BURNHAM, C. R.: Cytogenetic studies of a translocation between chromosomes 1 and 7 in maize. Genetics **33**, 5–21 (1948).

BURNHAM, C. R.: Chromosome segregation in translocations involving chromosome 6 in maize. Genetics **35**, 446–481 (1950).

BURNHAM, C. R.: Chromosomal interchanges in plants. Bot. Rev. **22**, 419–552 (1956).

BURNHAM, C. R.: Discussions in cytogenetics. Minneapolis: Burgess Publ. Comp. 1962.

BURNHAM, C. R., STOUT, J. T., WEINHEIMER, W. H., KOWLES, R. V., PHILLIPS, R. L.: Chromosome pairing in maize. Genetics **71**, 111–126 (1972).

BUSS, M. E., HENDERSON, S. A.: Induced bivalent interlocking and the course of meiotic chromosome synapsis. Nature New Biol. **234**, 243–246 (1971).

CALLAN, H. G., MONTALENTI, G.: Chiasma interference in mosquitoes. J. Genet. **48**, 119–134 (1947).

CAMENZIND, R., NICKLAS, R. B.: The non-random chromosome segregation in spermatocytes of *Gryllotalpa hexadactyla*. Chromosoma **24**, 324–335 (1968).

CARTER, T. C., FALCONER, D. S.: Stocks for detecting linkage in the mouse, and the theory of their design. J. Genet. **50**, 307–323 (1951).

CHAUHAN, K. P. S., ABEL, W. O.: Evidence for the association of homologous chromosomes during premeiotic stages in Impatiens and Salvia. Chromosoma **25**, 297–302 (1968).

COMINGS, D. E., OKADA, T. A.: Triple chromosome pairing in triploid chickens. Nature **231**, 119–121 (1971).

DARLINGTON, C. D.: The origin and behaviour of chiasmata, VIII. *Secale cereale* (*n*, 8). Cytologia **4**, 444–452 (1933).

DARLINGTON, C. D.: The origin and behaviour of chiasmata, VII. *Zea mays*. Z.I.A.V. **67**, 96–114 (1934).

DARLINGTON, C. D.: Cytology. Part I: Recent advances in cytology, 1937. Part II: Recent advances in cytology, 1937–1964. London: J. and A. Churchill Ltd. 1965.

DAVIES, E. D. G., JONES, G. H.: Chiasma variation and control in pollen mother cells and embryo-sac mother cells of rye. Genet. Res. **23**, 185–190 (1974).

DOUGLAS, L. T.: Meiosis III: The elastica, and Markov processes as models for non-random segregation from associated non-homologues in Drosophila. Genetica **39**, 289–328 (1968a).

DOUGLAS, L. T.: Meiosis IV: Segregation from interchange multivalents as a Markov process. Genetica **39**, 429–455 (1968b).

DOUGLAS, L. T.: Meiosis V: Matric and path coefficient solutions of tri- and quadrivalents. Genetica **39**, 456–496 (1968c).

DOUGLAS, L. T.: Meiosis VII: "Detorsive bending" as a basis for geometric shapes of late prophase bivalents. Genetica **41**, 231–256 (1970).

DOUGLAS, L. T.: Meiosis VIII. Segregation matrix methods applied to meiotic drive in Drosophila melanogaster males with an sc^4–sc^8 inverted X. Genetica **42**, 104–128 (1971).

DOVER, G. A., RILEY, R.: Variation at two loci affecting homoeologous meiotic chromosome pairing in *Triticum aestivum* × *Aegilops mutica* hybrids. Nature New Biol. **235**, 61–62 (1972a).

DOVER, G. A., RILEY, R.: Prevention of pairing of homoeologous meiotic chromosomes of wheat by an activity of supernumerary chromosomes of Aegilops. Nature **240**, 159–161 (1972b).

DOVER, G. A., RILEY, R.: The effect of spindle inhibitors applied before meiosis on meiotic chromosome pairing. J. Cell Sci. **12**, 143–161 (1973).

DOYLE, G. G.: Preferential pairing in structural heterozygotes of *Zea mays*. Genetics **48**, 1011–1027 (1963).

DRISCOLL, C. J., DARVEY, N. L.: Chromosome pairing: effect of colchicine on an isochromosome. Science **169**, 290–291 (1970).

DRISCOLL, C. J., DARVEY, N. L., BARBER, H. N.: Effect of colchicine on meiosis of hexaploid wheat. Nature **216**, 687–688 (1967).

DYER, A. F.: Heterochromatin in American and Japanese species of Trillium, III. Chiasma frequency and distribution and the effect on it of heterochromatin. Cytologia **29**, 263–279 (1964).

EINSET, J.: Chromosome length in relation to transmission frequency of maize trisomes. Genetics **28**, 349–364 (1943).

EKBERG, I.: Cytogenetic studies of three paracentric inversions in barley. Hereditas **76**, 1–30 (1974).

ELLIOTT, C. G.: Environmental effects on the distribution of chiasmata among nuclei and bivalents and correlation between bivalents. Heredity **12**, 429–439 (1958).

EMERSON, S.: Linkage and recombination at the chromosomal level. Genetic Organization I: (E. W. CASPARI and A. W. RAVIN, eds.), 267–360; 1969.

ENDRIZZI, J. E., KOHEL, R. J.: Use of telosomes in mapping three chromosomes of cotton. Genetics **54**, 535–550 (1966).

EVANS, G. M., MACEFIELD, A. J.: The effect of B-chromosomes on homoeologous pairing in species hybrids I. *Lolium temulentum* × *Lolium perenne*. Chromosoma **41**, 63–73 (1973).

FELDMAN, M.: The effect of chromosomes 5B, 5D, and 5A on chromosomal pairing in *Triticum aestivum*. Proc. Nat. Acad. Sci. **55**, 1447–1453 (1966).

FELDMAN, M., MELLO-SAMPAYO, T., SEARS, E. R.: Somatic association in *Triticum aestivum*. Proc. Nat. Acad. Sci. **56**, 1192–1199 (1966).

FIIL, A., MOENS, P. B.: The development, structure and function of modified synaptonemal complexes in mosquito oocytes. Chromosoma **41**, 37–62 (1973).

FOGWILL, M.: Differences in crossing over and chromosome size in the sex cells of Lillium and Fritillaria. Chromosoma **9**, 493–504 (1958).

FORD, C. E.: Meiosis in mammals. In: BENIRSCHKE, K.: Comparative mammalian cytogenetics. pp. 91–166. Berlin-Heidelberg-New York: Springer Verlag 1969.

FORER, A., KOCH, C.: Influence of autosome movements and of sex-chromosome movements on sex-chromosome segregation in Crane fly spermatocytes. Chromosoma **40**, 417–442 (1973).

FOX, D. P.: The control of chiasma distribution in the locust, *Schistocerca gregaria* (Forskål). Chromosoma **43**, 289–328 (1973).

FU, T. K., SEARS, E. R.: The relationship between chiasmata and crossing over in *Triticum aestivum*. Genetics **75**, 231–246 (1973).

GAIRDNER, A. E., DARLINGTON, C. D.: Ring formation in diploid and polyploid *Campanula persicifolia*. Genetica **13**, 113–150 (1931).

GAUL, H.: A critical survey of genome analysis. Proc. I Internat. Wheat Genet. Symp., Winnipeg, 194–206 (1959).

GILES, N. H.: The origin of iso-chromosomes at meiosis. Genetics **28**, 512–524 (1943).

GILES, N. H.: A pericentric inversion in Gasteria resulting in apparent iso-chromosomes at meiosis. Proc. Nat. Acad. Sci. **30**, 1–51 (1944).

GILLES, A., RANDOLPH, L. F.: Reduction of quadrivalent frequency in autotetraploid maize during a period of ten years. Am. J. Bot. **38**, 12–17 (1951).

GRELL, R. F.: Non-random assortment of non-homologous chromosomes in *Drosophila melanogaster*. Genetics **44**, 421–435 (1959).

GRELL, R. F.: Chromosome size at distributive pairing in *Drosophila melanogaster* females. Genetics **50**, 151–166 (1964).

GRELL, R. F.: Pairing at the chromosomal level. J. Cellular Physiol. **70** (Suppl. 1), 119–146 (1967).

GRELL, R. F., BANK, H., GASSNER, G.: Meiotic exchange without the synaptinemal complex. Nature New Biol. **240**, 155–157 (1972).

GRELL, R. F., DAY, J. W.: Chromosome pairing in the oogonial cells of *Drosophila melanogaster*. Chromosoma **31**, 434–445 (1970).

GUSTAFSSON, Å., HÅKANSSON, A.: Meiosis in some Rosa-hybrids. Botaniska Not. 331–343 (1942).

HALDANE, J. B. S.: The combination of linkage values, and the calculation of distance between the loci of linked factors. J. Genet. **8**, 299–309 (1919).

HALDANE, J. B. S.: The cytological basis of genetical interference. Cytologia **3**, 54–65 (1931).

HARTE, C.: Die Variabilität der Chiasmenbildung bei *Paeonia tenuifolia*. Chromosoma **8**, 152–182 (1956).

HEARNE, E. M., HUSKINS, C. L.: Chromosome pairing in *Melanoplus femur-rubrum*. Cytologia **6**, 123–147 (1935).

HEEMERT, C. VAN: Meiotic disjunction, sex-determination and embryonic lethality in an X-linked "simple" translocation of the onion fly, *Hylemya antiqua* (Meigen). Chromosoma **47**, 45–60 (1974).

HENDERSON, S. A.: Chiasma distribution at diplotene in a locust. Heredity **18**, 173–190 (1963).

HENDERSON, S. A.: Chiasma localisation and incomplete pairing. Chromosomes Today **2**, 56–60 (1969).

HENDERSON, S. A., KOCH, C. A.: Co-orientation stability by physical tension: a demonstration with experimentally interlocked bivalents. Chromosoma **29**, 207–216 (1970).

HINTON, C. W.: The effects of heterozygous autosomal translocations on recombination in the X-chromosome of *Drosophila melanogaster*. Genetics **51**, 971–982 (1965).

HULTÉN, M.: Chiasma distribution at diakinesis in the normal human male. Hereditas **76**, 55–78 (1974).

HUSKINS, C. L., NEWCOMBE, H. B.: An analysis of chiasma pairs showing chromatid interference in *Trillium erectum* L. Genetics **26**, 101–127 (1941).

JAGIELLO, G.: Meiosis in female mammals. Chromosomes Today V. (P. L. PEARSON and K. LEWIS, eds.). In press, 1975.

JAIN, H. K., BASAK, S. L.: Genetic interpretation of chiasmata in Delphinium. Genetics **48**, 329–339 (1963).

JOHN, B., HENDERSON, S. A.: Asynapsis and polyploidy in *Schistocerca paranensis*. Chromosoma **13**, 111–147 (1962).

JOHN, B., LEWIS, K. R.: The meiotic system. Protoplasmatologia VI F1. Wien-New York: Springer Verlag, 1965.

JONES, G. H.: The control of chiasma distribution in rye. Chromosoma **22**, 69–90 (1967).

JONES, G. H.: Light and electron microscope studies of chromosome pairing in relation to chiasma localisation in *Stethophyma grossum* (Orthoptera: Acrididae). Chromosoma **42**, 145–162 (1973).

JONES, G. H.: Correlated components of chiasma variation and the control of chiasma distribution in rye. Heredity **32**, 375–387 (1974).

JONES, R. N., REES, H.: Genotypic control of chromosome behaviour in rye XI. The influence of B-chromosomes on meiosis. Heredity **22**, 333–347 (1967).

JUDD, B. H.: The structure of intralocus duplication and deficiency chromosomes produced by recombination in *Drosophila melanogaster*, with evidence for polarised pairing. Genetics **49**, 253–265 (1964).

JUDD, B. H.: Chromosome pairing and recombination in *Drosophila melanogaster*. Genetics **52**, 1229–1233 (1965).

KAYANO, H.: Chiasma studies in structural hybrids. I. Heteromorphic bivalent in *Lilium callosum*. Nucleus **2**, 47–50 (1959).

KAYANO, H.: Chiasma studies in structural hybrids IV. Crossing-over in *Disporum sessile*. Cytologia **25**, 468–475 (1960).

KELMAN, M.: The forces influencing chromosome pairing in *Drosophila melanogaster*. Am. Naturalist **79**, 567–570 (1945).

KEMPANNA, C., RILEY, R.: Secondary association between genetically equivalent bivalents. Heredity **19**, 289–299 (1964).

KHOSHOO, T. N., MUKHERJEE, I.: A translocation heterozygote in garden Canna. Genetica **37**, 255–258 (1966).

KIMBER, G., RILEY, R.: Haploid angiosperms. Bot. Review **29**, 480–531 (1963).

KING, R. C.: The meiotic behavior of the Drosophila oocyte. Internat. Rev. Cytol. **28**, 125–168 (1970).

KISHIKAWA, H.: Cytogenetic studies of B-chromosomes in rye, *Secale cereale* L., in Japan. Agric. Bull. Saga Univ. **21**, 1–81 (1965).

KISHIKAWA, H.: Cytological study on triploid rye with four accessory chromosomes. Jap. J. Genet. **41**, 427–437 (1966).

KOLLER, P. C.: The origin and behaviour of chiasmata. XI. Dasyurus and Sarcophilus. Cytologia **7**, 82–103 (1936).

KOSAMBI, D. D.: The estimation of map distance from recombination values. Ann. Eugen. **12**, 172–175 (1944).

KÜNZEL, G.: Differenziertes Bindungsverhalten der Meiose-Chromosomen innerhalb der Ähre einer röntgeninduzierten Gerstenmutante mit reziproker Translokation in heterozygotem Zustand. Die Kulturpflanze **11**, 517–534 (1963).

LACADENA, J. R., PUERTAS, M. J.: Secondary association of bivalents in an allohexaploid *Aegilops triaristata* Willd. 6x. Genét. Ibér. **21**, 191–209 (1969).

LA COUR, L. F., WELLS, B.: Meiotic prophase in anthers of asynaptic wheat. Chromosoma **29**, 419–427 (1970).

LA COUR, L. F., WELLS, B.: Abnormalities in synaptonemal complexes in pollen mother cells of a lily hybrid. Chromosoma **42**, 137–144 (1973).

LAMM, R.: Cytological studies on inbred rye. Hereditas **22**, 217–240 (1936).

LAMMERTS, W. E.: On the nature of chromosome association in *N. tabacum* haploids. Cytologia **6**, 38–50 (1934).

LAWRENCE, C. W.: Genotypic control of chromosome behaviour in rye. VI. Selection for disjunction frequency. Heredity **12**, 127–131 (1958).

LÉONARD, A.: Radiation-induced translocations in spermatogonia of mice. Mutation Res. **11**, 71–88 (1971).

LÉONARD, A., DEKNUDT, G.: Chromosome rearrangements induced in the mouse by embryonic X-irradiation. I. Pronuclear stage. Mutation Res. **4**, 689–697 (1967).

LEVAN, A.: Zytologische Studien an *Allium schoenoprasum*. Hereditas **22**, 1–128 (1936).

LEVAN, A.: The effect of colchicine on meiosis in Allium. Hereditas **25**, 9–26 (1939).

LEVAN, A.: Meiosis in *Allium porrum*, a tetraploid species with chiasma localization. Hereditas **26**, 454–462 (1940).

LEVAN, A.: Studies on the meiotic mechanisms of haploid rye. Hereditas **28**, 177–211 (1942).

LINNERT, G.: Untersuchungen an hemiploiden Nachkommen Autotetraploider 1. Der Verteilungsmodus einer reziproken Translokation und die daraus folgende Erhöhung der Heterozygotenfrequenz in der Nachkommenschaft von Duplex-Heterozygoten. Z. Vererbungslehre **93**, 389–398 (1962).

LU, B. C., RAJU, N. B.: Meiosis in Coprinus II. Chromosome pairing and the lampbrush diplotene stage of meiotic prophase. Chromosoma **29**, 305–316 (1970).

MAEDA, T.: On the configurations of the gemini in the pollen mother cells of *Vicia faba* L. Mem. Coll. Sci. Kyoto Imp. Univ. **5**, 125–137 (1930).

MAGOON, M. L., KHANNA, K. R.: Haploids. Caryologia **16**, 191–235 (1963).

MAGUIRE, M. P.: The relationship of cross over frequency to synaptic extent at pachytene in maize. Genetics **51**, 23–40 (1965).

MAGUIRE, M. P.: The relationship of crossing-over to chromosome synapsis in a short paracentric inversion. Genetics **53**, 1071–1077 (1966).

MAGUIRE, M. P.: Evidence for homologous pairing of chromosomes prior to meiotic prophase in maize. Chromosoma **21**, 221–231 (1967).

MAGUIRE, M. P.: The temporal sequence of synaptic initiation, crossing-over and synaptic completion. Genetics **70**, 353–370 (1972).

MATHER, K.: Reductional and equational separation of the chromosomes in bivalents and multivalents. J. Genet. **30**, 53–78 (1935).

MATHER, K.: Competition between bivalents during chiasma formation. Proc. Royal Soc. B. **120**, 208–227 (1936).

MATHER, K.: The determination of position in crossing-over. II. The chromosome length-chiasma frequency relation. Cytologia (Fujii jub. vol.), 514–526 (1937).

MATHER, K.: Crossing-over. Biol. Rev. Cambr. **13**, 252–292 (1938).

MATHER, K.: Competition for chiasmata in diploid and trisomic maize. Chromosoma **1**, 119–129 (1939).

MATHER, K.: The determination of position in crossing-over. III. The evidence of metaphase chiasmata. J. Genetics **39**, 205–223 (1940).

MATHER, K.: The genetical structure of populations. Symp. Soc. Exp. Biol. **7**, 66–95 (1953).

MATHER, K.: Statistical analysis in biology. University Paperbacks, London: Methuen 1965.

MATHER, K., LAMM, R.: The negative correlation of chiasma frequencies. Hereditas **20**, 65–70 (1935).

McCLINTOCK, B.: The association of non-homologous parts of chromosomes in the mid-prophase of meiosis in *Zea mays*. Zeitschr. Zellforsch. u. mikr. Anat. **19**, 191–237 (1933).

McCOLLUM, G. D.: Comparative studies of chromosome pairing in natural and induced tetraploid Dactylis. Chromosoma **9**, 571–605 (1958).

MELLO-SAMPAYO, T.: Somatic association of telocentric chromosomes carrying homologous centromeres in common wheat. T.A.G. **43**, 174–181 (1973).

MENZEL, M. Y., PRICE, J. M.: Fine structure of synapsed chromosomes in F_1 *Lycopersicum esculentum—Solanum lycopersicoides* and its parents. Am. J. Bot. **53**, 1079–1086 (1966).

MOENS, P. B.: The time of chromosome pairing at meiosis. Genetics **54**, 349 (1966).

MOENS, P. B.: The structure and function of the synaptonemal complex in *Lilium longiflorum* sporocytes. Chromosoma **23**, 418–451 (1968a).

MOENS, P.: Fine structure aspects of chromosome pairing at meiotic prophase. Genetics **60**, 205–206 (1968b) (Abstr.).

MOENS, P.: Chromosome pairing, chiasmata and crossing over in asynaptic tomato mutants. Canad. J. Genet. Cytol. **10**, 769–770 (1968c) (Abstr.).

MOENS, P. B.: The fine structure of meiotic chromosome polarization and pairing in *Locusta migratoria* spermatocytes. Chromosoma **28**, 1–25 (1969).

MORGAN, T. H., BRIDGES, C. B., SCHULTZ, J.: Constitution of the germinal material in relation to heredity. Carnegie Inst. Wash. Yearbook **32**, 298–302 (1933).

MORRISON, J. W., RAJHATHY, T.: Chromosome behavior in autotetraploid cereals and grasses. Chromosoma **11**, 297–309 (1960a).

MORRISON, J. W., RAJHATHY, T.: Frequency of quadrivalents in autotetraploid plants. Nature **187**, 528–530 (1960b).

MOSES, M. J.: Structure and function of the synaptonemal complex. Genetics Suppl. **61** (2), 41–51 (1969).

NEWCOMBE, H. B.: Chiasma interference in *Trillium erectum*. Genetics **26**, 128–136 (1941).

NICKLAS, R. B.: Recurrent pole-to-pole movements of the sex chromosome during prometaphase I in *Melanoplus differentialis* spermatocytes. Chromosoma **12**, 97–115 (1961).

NICKLAS, R. B.: Chromosome segregation: an explanation of kinetochore reorientation. Genetics **60**, 207–208 (1968).

NICKLAS, R. B.: Mitosis. Adv. Cell Biol. **2**, 225–297 (1971).

NODA, S.: Chiasma studies in structural hybrids VII. Reciprocal translocation in *Scilla scilloides*. Cytologia **26**, 74–77 (1961).

NODA, S.: Achiasmate bivalent formation by parallel pairing in PMCs of *Fritillaria amabilis*. Bot. Mag. Tokyo **81**, 344–345 (1968).

NUR, U.: Synapsis and crossing over within a paracentric inversion in the grasshopper *Camnula pellucida*. Chromosoma **25**, 198–214 (1968).

OWEN, A. R. G.: The theory of genetic recombination. Adv. Genet. **3**, 117–157 (1950).

PASTOR, J. B., CALLAN, H. G.: Chiasma formation in spermatocytes and oocytes of the turbellarian *Dendrocoelum lacteum*. J. Genet. **50**, 449–454 (1952).

PÄTAU, K.: Cytologischer Nachweis einer positiven Interferenz über das Centromer. (Der Paarungskoeffizient. I.). Chromosoma **2**, 36–63 (1941).

PAYNE, F.: A study of the germ cells of *Gryllotalpa borealis* and *Gryllotalpa vulgaris*. Journ. Morph. **28**, 287–327 (1916).

PEACOCK, W. J., MIKLOS, G. L. G.: Meiotic drive in Drosophila: new interpretations of the segregation distorter and sex chromosome systems. Adv. Genet. **17**, 361–409 (1973).

PERRY, P. E., JONES, G. H.: Male and female meiosis in grasshoppers. I. *Stethophyma grossum*. Chromosoma **47**, 227–236 (1974).

PHILLIPS, L. L.: Cytogenetical evidence on the question of affinity in cotton. Heredity **19**, 21–26 (1964a).

PHILLIPS, L. L.: Segregation in new allopolyploids of Gossypium. V. Multivalent formation in New World × Asiatic and New World × Wild American hexaploids. Am. J. Bot. **51**, 324–329 (1964b).

POLANI, P. E.: Centromere localization at meiosis and the position of chiasmata in the male and female mouse. Chromosoma **36**, 343–374 (1972).

PRAKKEN, R.: Studies of asynapsis in rye. Hereditas **29**, 475–495 (1943).

PRICE, D. J.: Variation in chiasma frequency in *Cepaea nemoralis*. Heredity **32**, 211–217 (1974).

REES, H.: Genotypic control of chromosome behaviour in rye I. Inbred lines. Heredity **9**, 93–116 (1955).

REES, H.: Heterosis in chromosome behaviour. Proc. Roy Soc. B; **144**, 150–159 (1955).

REES, H.: Genotypic control of chromosome form and behaviour. Bot. Rev. **27**, 288–318 (1961).

REES, H., JONES, G. H.: Chromosome evolution in Lolium. Heredity **22**, 1–18 (1967).

REES, H., NAYLOR, B.: Developmental variation in chromosome behaviour. Heredity **15**, 17–27 (1960).

REES, H., SUN, S.: Chiasma frequency and the disjunction of interchange association in rye. Chromosoma **16**, 500–510 (1965).

RHOADES, M. M.: Preferential segregation in maize. In: (J. W. GOWEN, ed.) Heterosis, 66–80. Ames: Iowa State College Press 1952.

RHOADES, M. M.: Studies on the cytological basis of crossing over. Replication and recombination of genetic material. (W. J. PEACOCK and R. D. BROCK, eds.) Austral. Acad. Sci. **229–241** (1968).

RHOADES, M. M., DEMPSEY, E.: Cytogenetic studies of deficient-duplicate chromosomes derived from the inversion heterozygotes in maize. Am. J. Bot. **40**, 405–424 (1953).

RHOADES, M. M., DEMPSEY, E.: The effect of abnormal chromosome 10 on preferential segregation and crossing over-in maize. Genetics **53**, 989–1020 (1966).

RIEGER, R.: Inhomologenpaarung und Meioseablauf bei haploiden Formen von *Antirrhinum majus* L. Chromosoma **9**, 1–38 (1957).

RILEY, R.: The diploidisation of polyploid wheat. Heredity **15**, 407–429 (1960).

RILEY, R., CHAPMAN, V.: Haploids and polyhaploids in Aegilops and Triticum. Heredity **11**, 195–207 (1957).

RILEY, R., CHAPMAN, V.: The genetic control of cytologically diploid behaviour of hexaploid wheat. Nature **182**, 713–715 (1958).

RILEY, R., CHAPMAN, V.: Estimates of the homoeology of wheat chromosomes by measurements of differential affinity at meiosis. Chromosome manipulations and plant genetics. Suppl. Heredity **20**, 46–58 (1966).

RILEY, R., CHAPMAN, V., BELFIELD, A. M.: Induced mutation affecting the control of meiotic chromosome pairing in *Triticum aestivum*. Nature **211**, 368–369 (1966).

ROBERTS, P. A.: Screening for X-ray-induced cross over suppressors in *Drosophila melanogaster*: prevalence and effectiveness of translocations. Genetics **65**, 429–448 (1970).

ROBERTS, P. A.: Differences in synaptic affinity of chromosome arms of *Drosophila melanogaster* revealed by differential sensitivity to translocation heterozygosity. Genetics **71**, 401–415 (1972).

ROHLOFF, H.: Die Spermatocytenteilungen der Tipuliden. IV Mitteilung. Analyse der Orientierung röntgenstrahleninduzierter Quadrivalente bei *Pales ferruginea*. Dissertation Eberhard-Karls-Universität Tübingen, 1970.

ROWLANDS, D. G.: The control of chiasma frequency in *Vicia faba* L. Chromosoma **9**, 176–184 (1958).

SANDLER, L., NOVITSKI, E.: Meiotic drive as an evolutionary force. Am. Natural. **91**, 105–110 (1957).

SANNOMIYA, M.: Chiasma studies in structural hybrids. X. Further studies in *Acrida lata*. Jap. J. Genet. **43**, 103–108 (1968).

SATINA, S., BLAKESLEE, A. F.: Chromosome behavior in triploid *Datura stramonium* I. The male gametophyte. Am. J. Bot. **24**, 518–527 (1937).

SAX, K.: Variation in chiasma frequencies in Secale, Vicia and Tradescantia. Cytologia **6**, 289–293 (1935).

SCHRADER, F.: Mitosis. New York; Columbia Univ. Press 1953.

SCHULTZ, J., HUNGERFORD, D. A.: Characteristics of pairing in the salivary gland chromosomes of *Drosophila melanogaster*. Genetics **38**, 689 (1953).

SCHULTZ, J., REDFIELD, H.: Interchromosomal effects on crossing-over in Drosophila. Cold Spring Harbor Symp. Quant. Biol. **16**, 175–197 (1951).

SEARLE, A. G., FORD, C. E., BEECHEY, C. V.: Meiotic disjunction in mouse translocations and the determination of centromere position. Genet. Res. **18**, 215–235 (1971).

SEARS, E. R.: Chromosome engineering in wheat. Stadler Symp. **4**, 23–38 (1972a).

SEARS, E. R.: Reduced proximal crossing-over in telocentric chromosomes of wheat. Genét. Ibér. **24**, 233–239 (1972b).

SEARS, E. R., OKAMOTO, M.: Intergenomic chromosome relationships of non-homologous chromosomes in wheat. Proc. 10th Int. Congr. Genet. **2**, 258–259 (1958).

SFICAS, A. G.: Statistical analysis of chromosome pairing in interspecific hybrids. I. The probability distributions. Genetics **47**, 1163–1170 (1962).

SFICAS, A. G.: Statistical analysis of chromosome distribution to the poles in interspecific hybrids with variable chromosome pairing. Genet. Res. **4**, 266–275 (1963).

SFICAS, A. G., GERSTEL, D. U.: Statistical analysis of chromosome pairing in interspecific hybrids. II. Applications to some Nicotiana hybrids. Genetics **47**, 1171–1185 (1962).

SHAH, S. S.: Studies on a triploid, a tetrasomic triploid and a trisomic plant of *Dactylis glomerata*. Chromosoma **15**, 469–477 (1964).

SHAVER, D. L.: A study of meiosis in perennial teosinte, in tetraploid maize and in their tetraploid hybrid. Caryologia **15**, 43–57 (1962).

SHAVER, D. L.: The effect of structural heterozygosity on the degree of preferential pairing in allotetraploids of Zea. Genetics **48**, 515–524 (1963).

SIMCHEN, G., STAMBERG, J.: Genetic control of recombination in *Schizophyllum commune*: specific and independent regulation of adjacent and non-adjacent chromosomal regions. Heredity **24**, 369–381 (1969).

SMITH, S. G.: Polarization and progression in pairing. II. Premeiotic orientation and the initiation of pairing. Canad. J. Research D **20**, 221–229 (1942).

SNEDECOR, G. W., COCHRAN, W. G.: Statistical methods. Ames, Iowa, USA. The Iowa State Univ. Press, 6th Ed. 1967.

STACK, S.: Premeiotic changes in *Ornithogalum virens*. Bull. Torr. Bot. Club **98**, 207–214 (1971).

STAM, P.: Personal communication, 1974.

STEINITZ-SEARS, L. M.: Cytogenetic studies bearing on the nature of the centromere. Genetics Today. Proc. XI Int. Congr. Genet. **I**, 123 (1963).

STERN, H., HOTTA, Y.: Biochemical controls of meiosis. Ann. Rev. Genet. **7**, 37–66 (1973).

SUN, S., REES, H.: Genotypic control of chromosome behaviour in rye IX. The effect of selection on the disjunction frequency of interchange associations. Heredity **22**, 249–254 (1967).

SVED, J. A.: Telomere attachment of chromosomes. Some genetical and cytological consequences. Genetics **53**, 747–756 (1966).

SWAMINATHAN, M. S., MURTY, B. R.: Aspects of asynapsis in plants. I. Random and non-random chromosome associations. Genetics **44**, 1271–1280 (1959).

SYBENGA, J.: Inbreeding effects in rye. Z. Vererbungslehre **89**, 338–354 (1958).

SYBENGA, J.: Non-random distribution of chiasmata in rye, Crotalaria and coffee. Chromosoma **11**, 441–455 (1960).

SYBENGA, J.: The quantitative analysis of chromosome pairing and chiasma formation based on the relative frequencies of M I configurations. I. Introduction: Normal diploids. Genetica **36**, 243–252 (1965a).

SYBENGA, J.: The quantitative analysis of chromosome pairing and chiasma formation based on the relative frequencies of M I configurations. II. Primary trisomics. Genetica **36**, 339–350 (1965b).

SYBENGA, J.: The quantitative analysis of chromosome pairing and chiasma formation based on the relative frequencies of M I configurations. III. Telocentric trisomics. Genetica **36**, 351–361 (1965c).

SYBENGA, J.: Reciprocal translocations and preferential pairing in autotetraploid rye. Chromosomes Today. (C. D. DARLINGTON and K. R. LEWIS, eds.) Vol. **1**, 66–70 (1966a).

SYBENGA, J.: The zygomere as hypothetical unit of chromosome pairing initiation. Genetica **37**, 186–198 (1966b).

SYBENGA, J.: The quantitative analysis of chromosome pairing and chiasma formation based on the relative frequencies of M I configurations. IV. Interchange heterozygotes. Genetica **37**, 199–206 (1966c).

SYBENGA, J.: Interchromosome effects on chiasma frequencies in rye. Genetica **38**, 171–183 (1967a).

SYBENGA, J.: The quantitative analysis of chromosome pairing and chiasma formation based on the relative frequencies of M I configurations. V. Interchange trisomics. Genetica **37**, 481–510 (1967b).

SYBENGA, J.: Orientation of interchange multiples in *Secale cereale*. Heredity **23**, 73–79 (1968).

SYBENGA, J.: Allopolyploidization of autopolyploids. I. Possibilities and limitations. Euphytica **18**, 355–371 (1969).

SYBENGA, J.: The calculation of the map length of interchange segments from meiotic configuration frequencies. Genetica **41**, 101–110 (1970a).

SYBENGA, J.: Simultaneous negative and positive chiasma interference across the break point in interchange heterozygotes. Genetica **41**, 209–230 (1970b).

SYBENGA, J.: General Cytogenetics. Amsterdam-London-New York: North-Holland/American Elsevier 1972a.

SYBENGA, J.: Localisation of intraspecific variation in meiotic chromosome pairing. Chromosomes Today **3**, 104–109 (1972b). Suppl. Heredity.

SYBENGA, J.: Chromosome associated control of meiotic pairing differentiation. Variation within *Secale cereale*. Chromosoma **39**, 351–360 (1972c).

SYBENGA, J.: The effect of reciprocal translocations on segregation and multivalent formation in autotetraploids of rye, *Secale cereale*. Genetica **44**, 270–282 (1973a).

SYBENGA, J.: Allopolyploidization of autopolyploids 2. Manipulation of the chromosome pairing system. Euphytica **22**, 433–444 (1973b).

SYBENGA, J.: The quantitative analysis of chromosome pairing and chiasma formation based on the relative frequencies of M I configurations. VII. Autotetraploids. Chromosoma **50**, 211–222 (1975).

SYBENGA, J., VERHAAR, H.: Meiotic behaviour of combinations of isochromosomes and telocentrics in rye. Chromosoma, in press, 1975.

SYBENGA, J., DE VRIES, J. M.: Chromosome pairing and chiasma formation in polysomic B-chromosomes in rye, *Secale cereale*. Biol. Zentrbl. **91**, 181–192 (1972).

SYBENGA, J., WILMS, H. J., MULDER, A. D.: Genetic length and interference in telo-substituted interchange heterozygotes of rye. Chromosoma **42**, 95–104 (1973).

TIMMIS, J. N., REES, H.: A pairing restriction at pachytene upon multivalent formation in autotetraploids. Heredity **26**, 269–275 (1971).

VARDI, A., DOVER, G. A.: The effect of B chromosomes on meiotic and pre-meiotic spindles and chromosome pairing in Triticum/Aegilops hybrids. Chromosoma **38**, 367–385 (1972).

VED BRAT, S.: Genetic systems in Allium II. Sex differences in meiosis. Chromosomes Today **I** (DARLINGTON and LEWIS, eds.), 31–40 (1966).

VENKATESWARLU, J.: Meiosis in autotetraploid maize (*Zea mays*). Ph. D. Thesis, Cambridge 1950. (Cf. SVED, 1966.)

VOSA, C. G.: Two-track heredity: differentiation of male and female meiosis in Tulbaghia. Caryologia **25**, 275–281 (1972).

WALLACE, M. E.: Affinity: evidence from crossing inbred lines of mice. Heredity **16**, 1–23 (1961).

WALTERS, M. S.: Evidence on the time of chromosome pairing from the preleptotene spiral stage in *Lillium longiflorum* "Croft". Chromosoma **29**, 375–418 (1970).

WALTERS, M. S., GERSTEL, D. U.: A cytological investigation of a tetraploid *Rhoeo discolor*. Am. J. Bot. **35**, 141–150 (1948).

WATSON, J. D., CALLAN, H. G.: The form of bivalent chromosomes in newt oocytes at first metaphase of meiosis. Quart. J. Micro. Sci. **104**, 281–295 (1963).

WESTERGAARD, M., VON WETTSTEIN, D.: The synaptinemal complex. Ann. Rev. Genetics **6**, (3036) 71–110 (1972).

WESTERMAN, M.: The effect of X-irradiation on male meiosis in *Schistocerca gregaria* (Forskål) I. Chiasma frequency response. Chromosoma **22**, 401–416 (1967).

WHITE, M. J. D.: Animal cytology and evolution. 2nd. Ed. Cambridge: Cambridge Univ. Press 1954.

WHITE, M. J. D.: Animal cytology and evolution. 3rd. Ed. Cambridge: Cambridge University Press, 1973.

WILSON, J. Y.: Chiasma frequency in relation to temperature. Genetica **29**, 290–303 (1959).

WU, T. P., PI, C. P.: Reciprocal translocations induced by X-rays and thermoneutrons in wild sorghum (*S. purpureo-sericeum* Aschers and Schweinf.). Canad. J. Genet. Cytol. **10**, 152–160 (1968).

ZARCHI, Y., SIMCHEN, G., HILLEL, J., SCHAAP, T.: Chiasmata and the breeding system in wild populations of diploid wheats. Chromosoma **38**, 77–94 (1972).

ZEČEVIČ, L., PAUNOVIČ, D.: The effect of B-chromosomes on chiasma frequency in wild populations of rye. Chromosoma **27**, 198–200 (1969).

ZEN, S.: Chiasma studies in structural hybrids VI. Heteromorphic bivalent and reciprocal translocation in *Allium fistulosum*. Cytologia **26**, 67–73 (1961).

ZOHARY, D.: Chiasmata in a pericentric inversion in *Zea mays*. Genetics **40**, 874–877 (1955).

Author and Subject Index

Numbers in italics refer to pages where the relevant term is of special importance in the context, or where it is discussed in detail. Terms which occur throughout the book with high frequency (bivalent, chromosome, configuration, etc.) have not been indexed. Terms of an unspecific or general character (animals, heterozygous, etc.) have also not been included. For authors, and co-authors mentioned in the text, no page numbers of the list of references have been given. For co-authors not mentioned in the text the relevant page numbers of the list of references are given between brackets.

Sample copies as well as subscription and back-volume information available upon request

Please address:

Springer-Verlag
Werbeabteilung 4021
D-1000 Berlin 33
Heidelberger Platz 3

or

Springer-Verlag
New York Inc.
Promotion Department
175 Fifth Avenue
New York, NY 10010

TAG Theoretical and Applied Genetics

Internationale Zeitschrift für
Theoretische und Angewandte Genetik

Managing Editor: H. Stubbe

Breeding genetics, with the aid of chemistry and mathematics, has become considerably more fundamental and general. This development has moved from the genetics of the individual to that of the group and, in turn, to the study of the evolution and origin of domesticated species. Improved mathematical models, which can be quantitatively solved or simulated by the computer, allow the new science of molecular genetics to study gene-enzyme interaction and the regulation of inherited characteristics. TAG fills a vital need for detailed research reports in this field and serves as an international vehicle for the exchange of scientific information.

Chromosoma

Editorial Board: H. Bauer (Managing Editor), W. Beermann, J. G. Gall, R. B. Nicklas, H. Swift, J. H. Taylor

Chromosoma, founded in 1939, publishes original contributions concerning all aspects of nuclear and chromosome research. Current studies in this field range from those on protozoan chromosomes to research on the nuclei of higher organisms and frequently apply techniques and material from fields as diverse as molecular and population genetics. Pertinent biochemical and biophysical approaches to cytological problems are often included.

Springer-Verlag
Berlin
Heidelberg
New York

A Glossary of Genetics and Cytogenetics

Classical and Molecular

By R. Rieger, A. Michaelis,
M. M. Green

Third edition 1968. 90 figures
570 pages
ISBN 3-540-04316-0 Cloth DM 75,—
ISBN 0-387-04316-0 (North America)
Cloth $21.00

Distribution rights for U.K., the
Commonwealth, and the Traditional
British Market (excluding Canada):
Allen & Unwin Ltd., London

Prices are subject to change
without notice

From the reviews:
"What is expected of a good glossary
by the researcher, teacher and student?
The researcher, years out of school,
must often look up terms and defini-
tions which have faded from memory
or which have been revised, their
meaning narrowed or amplified. For
the teacher a glossary is indispensable
in keeping the general knowledge of
his rapidly growing discipline up-to-
date, now a more difficult task in
genetics than ever before. For the
student is should be an efficient aid to
his text. To serve all, a good glossary
should contain carefully selected
entries and concise definition of terms.
Last but not least, it should be easy
to find all terms related to the original
entry by a good cross-reference system.
This glossary eminently fulfills all
these requirements. In fact, it is more
than a glossary because the treatment
of the subjects is more encyclopedic
than one would expect of a technical
dictionary. It also describes concepts
and theories often supported by the
most relevant experimental data.
There have been two earlier editions
in German, but this English edition is
a completely new book. The subtitle
is well justified by the content which
covers both classical and molecular
genetics very comprehensively.
Because of this, "cytogenetics", might
have been omitted from the title,
since other branches of genetics are
equally well covered. . . ."

Canadian Journal of
Genetics and Cytology

Springer-Verlag
Berlin
Heidelberg
New York